Julia Verne
Living Translocality

Erdkundliches Wissen

Schriftenreihe
für Forschung und Praxis

Begründet von
Emil Meynen

Herausgegeben
von Martin Coy,
Anton Escher
und Thomas Krings

Band 150

Julia Verne

Living Translocality

Space, Culture and Economy in Contemporary
Swahili Trade

 Franz Steiner Verlag

Gedruckt mit freundlicher Unterstützung
der Deutschen Forschungsgemeinschaft

Umschlagabbildung: ‚Translocal Swahili Space',
Linoleumdruck von Theo Aalders

Bibliografische Information der Deutschen Nationalbibliothek
Die Deutsche Nationalbibliothek verzeichnet diese
Publikation in der Deutschen Nationalbibliografie;
detaillierte bibliografische Daten sind im Internet über
<http://dnb.d-nb.de> abrufbar.

ISBN 978-3-515-10094-6

Jede Verwertung des Werkes außerhalb der
Grenzen des Urheberrechtsgesetzes ist unzulässig
und strafbar. Dies gilt insbesondere für Übersetzung,
Nachdruck, Mikroverfilmung oder vergleichbare
Verfahren sowie für die Speicherung in Datenver-
arbeitungsanlagen.
© Franz Steiner Verlag, Stuttgart 2012
Gedruckt auf säurefreiem, alterungsbeständigem
Papier.
Druck: Laupp & Göbel, Nehren
Printed in Germany

CONTENTS

List of Figures ix
Acknowledgements xi

INTRODUCTION: MOBILITY AND THE GEOGRAPHIES OF SPACE 1

Mobility Studies in Geography: The Question of Space in Mobile
    Contexts 1
Swahili Studies: Swahili as Inherently Translocal? 5
Focusing on Translocality in the Swahili Context 9
Structuring the Text: Arrangements – Movements – Enmeshments 10

ARRANGEMENTS

EPISTEMOLOGY: A RELATIONAL APPROACH TO TRANSLOCALITY 15

Positioning Translocality 15
Shortcomings of the 'Network' 20
The Metaphor of the Rhizome 23
1st halt: Translocality Rearranged 31

METHODOLOGY: A MOBILE ETHNOGRAPHY OF TRANSLOCAL TRADING PRACTICES 33

Setting the Field through trade 33
The Gains of Ethnography 36
First Encounters with a Translocal Field 43
Mobility as Method 48
Doing Mobile Ethnographic Research 56
2nd halt: Researching Translocality 63
Outlook: Four Kinds of Swahili Trading Connections 64

MOVEMENTS

FINDING ONE'S WAY IN(TO) TRANSLOCAL CONNECTIONS: A TRADE JOURNEY THROUGH THE TANZANIAN HINTERLAND 71

The Selection and Acquisition of Goods 71
Passing Travelling Time 76
Facing ‚Africa'? Views from the Coast 81
The Presence of History: Sumbawanga, Mpanda and Tabora 84
Facing the Coast: Views from the Mainland 97
Family and Trade: Reflections on Culture and Economy (I) 99
3rd halt: Finding one's Way in(to) Translocal Connections 103

LIVING (UP TO) THE TRANSLOCAL IMAGINATION: ON AND IN-BETWEEN BUSINESS TRIPS TO DUBAI 105

Becoming a Business Traveller 106
Dubai: Business with and without Family 107
   A Sidenote on the Situation of Swahili in the United Arab Emirates 108
The Transportation of Goods across the Indian Ocean 117
Keeping Shops 120
Demanding or Bypassing Trips: Shopping in Zanzibar during
   the Month of Ramadhan 133
Consumption from a Swahili Perspective: Reflections on
   Culture and Economy (II) 138
4th halt: Living (up to) the Translocal Imagination 140

STICKING (TO) TRADING CONNECTIONS: OBJECT GEOGRAPHIES IN AND THROUGH THE HANDS OF WANNABE TRADERS 143

Prelude: For Following Objects 143
Migration and Development? Receiving Goods from the United
   Kingdom 144
The Everyday Life of *Masela*: Following 'Urban Sailors' and their
   Goods 147
Trading with Mobile Phones: Trading Culture beyond Traders 156
An Ideology of Trade: Reflections on Culture and Economy (III) 163
   A brief Excursion into the Political Present of Zanzibar 168
5th halt: Sticking (to) Trading Connections 171

SEARCHING HOME IN A TRANSLOCAL SPACE: ECONOMIC DIMENSIONS OF TRANSLOCAL CULTURAL PRACTICES 173

Economic Rationalities of Cultural Practices and their Material
    Effects: Reflections on Culture and Economy (IV) 173
Finding Someone Worth Marrying 174
Putting Swahiliness in 'the Bag': Material Exchange before the
    Wedding 181
The Choice of the Wedding Location 182
Wedding Celebrations in London and Mombasa 184
Making Home after the Wedding 194
6th halt: Searching Home in a Translocal Space 202

ENMESHMENTS

MOBILITY AND THE GEOGRAPHIES OF SPACE: CREATING A TRANSLOCAL SPACE THROUGH TRADE 207

Translocality as a Lived Experience 207
The Dialectics of Transgression and Situatedness 208
Translocal Spaces 212
Culture and Economy in a Translocal Space 214

LIVING TRANSLOCALITY: GROUNDING THEORETICAL CONCEPTS IN THE ACTUAL LIVES OF ACTUAL PEOPLE 217

From 'Transnational Social Spaces' to Translocal Space: Reflections
    on Inner Structures and the Role of Location 217
Relational Space: The Matter of Distance 226
Mobility and Cosmopolitanism: Living the Translocal Space 234

REFERENCES 241

LIST OF FIGURES

Fig. 1: Rhizome of *Cimicifuga racemosa* 24
Fig. 2: Illustration of relational thinking 27
Fig. 3: Location of Zanzibar and Pemba 44
Fig. 4: My first host family in Zanzibar 45
Fig. 5: Michenzani 45
Fig. 6: Traveller, disembarking from the ferry to Zanzibar 61
Fig. 7: 'The first Arab of Sumbawanga' telling his stories 61
Fig. 8: Four kinds of Swahili trading connections 65
Fig. 9: Travel route through the Tanzanian hinterland 73
Fig. 10: Travelling by train 77
Fig. 11: Meeting our local agent in Sumbawanga 83
Fig. 12: At the old coffee house 87
Fig. 13: Amongst grandchildren 87
Fig. 14: Behind the counter 89
Fig. 15: Sketch of Mpanda 90
Fig. 16: Madukani 92
Fig. 17: The first encounter with camels 92
Fig. 18: Emirates' advertisement in Dar es Salaam 106
Fig. 19: Map of Dubai 111
Fig. 20: Family meetings in Dubai 112
Fig. 21: Gold Plaza Hotel in Dubai 115
Fig. 22: Number and direction of dhows in the Western Indian Ocean 1946 118
Fig. 23: Kiponda 122
Fig. 24: Selling boys' clothes in Kiponda 122
Fig. 25: Map of Zanzibar town's main shopping areas 123
Fig. 26: Shops in Macontainer 124
Fig. 27: Trader's calculation 124
Fig. 28: Grid-pattern suburb of Kariakoo in 1969 125
Fig. 29: Kariakoo today (I) 126
Fig. 30: Recent building activities in Kariakoo 127
Fig. 31: Kariakoo today (II) 129
Fig. 32: Inside a home shop 131
Fig. 33: Juda's Shopping Mall 134
Fig. 34: Dressing up for Eid-il-fitr 137
Fig. 35: *Masela* in front of the warehouse 150
Fig. 36: Time-space-diary of Matar and Seif 152
Fig. 37: Meeting potential customers in Pemba 159
Fig. 38: Mobile phone shopping centre in Aggrey, Kariakoo 162
Fig. 39: Map of Mombasa 176
Fig. 40: Getting prepared for the wedding 185
Fig. 41: *Nikkah* 187
Fig. 42: The newly-wed couple on stage 187
Fig. 43: Staging an Arabian night 190
Fig. 44: Newly-wed couple and male relatives 193
Fig. 45: Posing for the camera 193

ACKNOWLEDGEMENTS

'The best way to travel is to feel' (Fernando Pessoa)

Before inviting anyone to follow me on this journey along contemporary Swahili trading connections I wish to thank my hosts and halting points who, in various ways, contributed to the highly mobile, very diverse, and yet extremely intense couple of years in which I was working on this subject.

In a certain sense it all began in Hamburg when I was about to finish school and my friend Lisa asked me to join her in taking private lessons in Kiswahili. Even though, at that time, I hardly knew what degree to choose she enlisted me and, by doing so, somehow set the direction for my regional focus. A couple of years later, when I was studying geography at the University of Bayreuth specialising in Development Geography and African Studies, it was Olivier Graefe who made me experience the fascination of methodological and epistemological questions, and by doing so, somehow planted the idea in my head that venturing into academia might be an enjoyable endeavour. I thus feel indebted to both of them for their companionship and early inspiration. These two strands have clearly remained very influential throughout my time as a PhD student and this book is derived from the attempt to bring them together.

In respect to Swahili studies, I am particularly grateful to Gerlind Scheckenbach, Said and Rahma Khamis, Gudrun Miehe, Sauda Barwani and Ridder Samson not only for deepening my interest in Kiswahili, Swahili history and way of life, but also for forming a kind of Swahili parenthood in Bayreuth as well as beyond. Included here are Aïsha, Jasmin, Seba, Clarissa and Paola, all of them being an invaluable part of this.

With regard to Geography, it was certainly my time as a Masters student at Royal Holloway in London that was key to making me feel a passion for cultural geography. I wish to thank Tim Cresswell, Felix Driver, David Gilbert, Tim Unwin and Katie Willis as well as the Landscape Surgery more generally for providing me with such a thrilling context to develop and sharpen my ideas and arguments! Above all, I owe a great deal to Phil Crang who inspired and supported me and even agreed to serve as an external examiner of my PhD in Bayreuth.

It was Detlef Müller-Mahn who not only offered me a great position at his chair but also provided me with an ideal institutional and social context for working on this project. I am extremely grateful for his support and encouragement, his interest in the topic, stimulating discussions, as well as the freedom and independence he granted me to adjust the orientation of our project according to my interests. I also wish to thank my friends and colleagues at the University of Bayreuth who have made this such a pleasant and homely place to do this work, with special thanks to Martin Doevenspeck and Jonathan Everts for always making time for a chat, to Michael Wegener for his impressive cartographic skills, and to Sarah Utecht, Benny Reichpietsch, Karenina Schröder and Theo Aalders for their additional help. Especially regarding the final steps that led to this publication I also want to thank Susanne Henkel and Harald Schmitt from the Franz Steiner Verlag for their friendly and efficient support.

At this point, I also wish to thank the DFG (German Research Foundation) for the generous financial support – first in the context of a collaborative research project (Sonderforschungsbereich SFB FK 560) and later through the funding of our project 'Mobility, Translocality and Trade: Imaginative Geographies among the Swahili' (MU 1034/9-1) – without which the extensive research in different places and on the routes in-between upon which this book is based wouldn't have been possible.

My deepest feeling of gratitude is to all those who have been there for me in various ways during my research – to Khelef and his family, Sabra and Nassor, Bi Shi and Mohamed Massoud as well as their children – first and foremost Eddiy – in Pemba, Zanzibar and Dar, Dira and Majid in Sumbawanga, Aunty Stara in Mpanda, Mohamed and Moza in Tabora, Salamuu and her family, as well as Ami Jayd and Aunty Mgeni in Dubai, Baba T Abdallah and Najla, Aunty Rahima and Uncle Naim and – above all – Ilhaam in London, Prof. Hyder and his family in Mombasa, Masoud and his family in Zanzibar and Muscat and to all the others who helped me find my way in and in-between these places for their hospitality and friendship. Without them letting me become part of their families this book would be an entirely different one. And, last but not least, I also wish to thank my own family, my parents - my father for his openness and support, when I decided against a career in law but embarked on a journey to less familiar terrains instead, and my mum for trying to share as much of the experiences as possible when visiting me wherever I went.

Finally, an almost overwhelming feeling of gratitude is to Markus, my husband, for his (com)passion and patience to accompany me on my sometimes messy and myriad lines of thought and helping me refine them, and for making even the most monotonous times of writing and formatting special and fun.

INTRODUCTION:
MOBILITY AND THE GEOGRAPHIES OF SPACE

MOBILITY STUDIES IN GEOGRAPHY: THE QUESTION OF SPACE IN MOBILE CONTEXTS

Born in Zanzibar, living in London, thinking about moving to Dubai; having one child working in the UK, another one residing in Toronto, while the third one is about to get married in Mombasa; regularly spending one's holiday along the East African coast, while passing the rest of the year in Europe; negotiating one's Omani belonging in narrations, through looks and material objects, even though never having been there; travelling across the Indian Ocean by taking goods from one place to another; distributing presents and making business when visiting friends and relatives spread all over the globe; being on the move and yet being at home; being dispersed and yet feeling closely related. At the same time disruptive and connective, transgressing as well as connecting places in distant parts of the world, mobility not only challenges the everyday lives of millions of people, it also challenges scientific understandings of society and culture. And it challenges geographers' conceptualisations of space.

The burgeoning of mobility as a core aspect in the social sciences and humanities that can be observed in the last two decades has had a strong impact on how researchers – especially geographers – look at place and space. Relational approaches to place as elaborated most explicitly by Doreen Massey (1994b) and Ash Amin (2002, 2004), the idea of 'translocalities' as developed by M.P. Smith in his book *Transnational Urbanism* (2001), or, more generally, the almost omnipresent emphasis on networks as fuelled most effectively by Manuel Castells (1996), are only some of the most prominent examples of the ways in which place has been (re)conceptualised in order to account for mobile conditions. Moreover, ideas on transnational spaces (Pries 1996, 2008), 'folded space' (Deleuze 1988, Serres & Latour 1995) or '-scapes' (Appadurai 1996) illustrate some of the most far-reaching attempts to incorporate mobility and its impact on distance into contemporary thinking about space. It is these theoretical reflections and the ways in which they are taken up and developed further in the context of the flourishing field of mobility studies (cf. Cresswell 2006, Sheller & Urry 2006) that initiated the idea for this book. From a geographic perspective, this book is thus motivated by a deep interest in dealing with the ways in which mobility influences experiences, constructions and conceptualisations of space and place.

When trying to delve deeper into the literature on mobility, one soon comes across a certain tendency to address this issue on a very abstract level, driven by the ambitious aim to formulate general assumptions on how mobility influences 'the world'. In parts heavily relying on the authors' own perceptions, the resulting statements generally rest on 'the West', and often seem to be in line with broader ideologies and political discourses concerning mobility.

For many people mobility stands for progress, liberty, freedom, transgression, thus a desirable central quality of modern life, with figures such as the flâneur, the globetrotter and explorer as predominantly male heroes (Pallasmaa 2008: 145, Sheller 2008: 258, Kaplan 2006). This rather romantic reading can also be discovered in some of the recent writings in the context of mobility studies. Many research projects are still built on an overly abstract celebration of travel and mobility or, as Canzler et al. put it, on a certain ‚mobility fetishism' (Canzler et al. 2008: 2), thus pursuing a rather narrow and elitist perspective, while neglecting broader structures and unequal power relations that so often seem to underpin contemporary forms of mobility (cf. Mitchell 1997). On the other hand, a bulk of work exists that is built on an extremely negative reading of mobility. Focusing on chaos, disorder and instability, in these cases, mobility is conceptualised as a central disruptive force in the contemporary world, disturbing, among other things, people's relation to place and space, and, in respect to migratory regimes, making space less controllable. Especially studies on mobility in the global south, generally marked by a strong focus on development, tend to be pervaded by a clear sedentary bias (cf. Verne & Doevenspeck 2012).

As a result of these often very extreme and one-sided positions, over the last years, numerous calls have been made for a stronger emphasis on the mundaneness and everyday dimensions of mobility in a variety of settings (e.g. Cresswell 2006, Sheller & Urry 2006, Smith & Katz 1993). Only by being attentive to different experiences of mobility as they come together in everyday lives, it would be possible to avoid and go beyond these generalising and dichotomising discourses. Thus, there clearly seems to be the need for ethnographic research that is able to provide a more nuanced and complex picture of how mobility effects conceptions of place and space. And, indeed, by browsing through the new publications on mobility, for example in Ashgate's book series 'Transport and Society', but also in most of the main geography journals, one can already find several examples of how this might be done.

Of special interest here is a book edited by Cresswell and Merriman on *Geographies of Mobilities* that explicitly aims at bringing to the fore the "qualitative exceptions, differences and experiences of movement" that, instead of being taken seriously were, at least in geographic accounts, so often relegated to footnotes and asides (Cresswell & Merriman 2010: 4). By structuring the book around practices, spaces and subjects the editors are able to take up and add to three strands of research that also seem to characterise the recent engagements of geographers with mobility, place and space more widely. Many geographers that can be considered part of the so-called 'mobility turn' have so far tended to either concentrate on particular practices, as for example walking (Middleton 2011), dancing (Cresswell 2006), flying (Adey 2010b) or driving (Laurier 2004), focus on very specific sites such as airports (Adey 2004, 2006, 2007) or roads (Merriman 2007), or deal with subjects that have a very obvious – though this does not necessarily mean less complex - relation to mobility (e.g. tramps, tourists, migrants and refugees). Although the references are not always that clear-cut, it can still be observed, that, while the examination of practices especially facilitates a deeper insight into mobility as concrete acts of movement, the analyses of spaces particularly emphasise mobility's material effects, and a close attention to subjects often serves extremely well to bring the diverse experiences of mobility to the fore. Moreover, in all three strands of research, a sensitive engagement with the discourses of mobility has come to play a crucial role, often combined with a historical perspective on the phenomenon under study. Thus, by dealing with the empirical reality of the act of moving, its discursive meaning as well as the ways in which mobility is a particular experience of the world, *Geographies of Mobilities* altogether succeeds in providing a more holistic understanding of mobility, as claimed by Cresswell in his earlier work in which he elaborates on these three 'relational moments' of mobility (Cresswell 2006, 2008). However, this recent work on different mobilities also makes clear that a strong focus on either particular practices, sites or subjects makes it very difficult to provide a holistic understanding of mobility in the sense of contextualising mobile practices, sites and subjects. By singling out a particular mobile practice, mobile subjects, or a specific site highly relevant to these practices, what often seems to be missing is their embedding into wider lifeworlds. So, this is exactly what this book tries to do: to get to grips with the manifold and often ambiguous experiences and forms of mobility as they actually take place in, and inform the everyday lives of, ordinary people, and – referring to Cresswell once again – to more closely examine how these people 'experience the world through motion' (Cresswell 2008: 131).

Most people would probably agree that mobility impacts on geographies of belonging, processes of inclusion and exclusion, and ideas of home. It also seems undeniable today that mobility connects places, plays a decisive role in the creation of spaces and leads to a different sense of being in the world. But while all these statements suggest a close connection between mobility, place and space they are not yet very precise, and in general it seems as if the geographical research on mobility is still somehow separate from the more theoretical reflections on how a 'mobile approach' effects conceptions of space and place. As it can be observed in the work referred to above, the former is rather concerned with concrete mobilities and how they are negotiated and experienced in space, whereas the latter often remains more or less devoid of any concrete subjects and experiences and instead talks more about possible spatial outcomes.

A term that well evokes the close connection between these two fields is translocality. By transcending and going beyond a locale, the term is seen to encompass concrete movements as well as its various material and imaginative outcomes. Through people's 'translocality' different places become connected, and new spaces are created. And, apart from referring to a particular phenomenon, the term also stands for a certain perspective – a perspective that foregrounds different forms of mobility in constructions of place and space (Freitag & von Oppen 2010b). Translocality not only turns the attention to concrete movements, but also alludes to the more abstract theoretical ideas, thus appearing as an ideal lens through which to bring the two fields closer together and approach the ways in which mobility effects constructions, experiences and conceptualisations of place and space. However, discussions of 'translocality' are so far also often characterised by a strong division between theoretical arguments on the one hand, and empirical examples on the other. The theoretical elaborations often tend to be cut off from the kinds of detailed, textured and often rather messy knowledges characteristic of ethnographic research. Following a common practice among many social scientists, they are generally first developed on a rather abstract level to later be applied to the empirical examples in a rather deductive manner. Seldom, these empirical insights are referred back to the theoretical assumptions, laying open tensions and ruptures between the two, and thus helping to refine and adapt dominant conceptions of translocality.

The approach this book pursues has grown out of a certain discomfort and dissatisfaction with this kind of 'theoretically guided research', and instead tries to promote a more interpretive engagement with 'translocality'. As Gadamer has pointed out, building on Heidegger's ideas on a hermeneutic circle (Heidegger 1927: 153), 'a person who is trying to understand is exposed to distraction from fore-meanings that are not borne out by the things themselves. Working out appropriate projections, anticipatory in nature, to be confirmed "by the things" themselves, is the constant task of understanding' (Gadamer 1975: 270). This should surely not be mistaken for a 'naïve empiricism', as it instead advocates to 'let oneself be guided by the things themselves' (Gadamer 1975: 269), remain more open towards their meanings, and enrich theoretical reflections accordingly. Also when following an interpretive approach, 'theoretical ideas are not created wholly anew in each study; [...] they are adopted from other, related studies, and, [instead of being applied and remaining more or less fixed, they are constantly] refined in the process' (Geertz 1973: 27). In this respect, it is the aim of this book to ground existing theoretical reflections on translocality in everyday practices and lived experiences, and to crossread and rethink the theoretical arguments on that basis. As Geertz emphasises in his seminal essay *Thick Description*, in interpretive science 'progress is marked less by a perfection of consensus than by a refinement of debate' (Geertz 1973: 29).

Does it, as some of the work on translocality suggests, due to intensive connections and entanglements between certain places, really not make any difference to some people in which of them they actually live? Does physical distance really not matter anymore to people living in a translocal context, so that geographers should actually abandon the concept of topographical space in favour of topology? Does high mobility between different places actually lead to a translocal sense of home or even make people feel to be at home in the world? Only an understanding of how translocality is actually lived, made meaningful and experienced is able to show how place and space are constructed in mobile settings, thus allowing us to critically engage with dominant conceptions of mobilities' effects on these two major geographical concepts.

SWAHILI STUDIES: SWAHILI AS INHERENTLY TRANSLOCAL?

'This is how people like us make a living: travel, trade and make our way in the world.' (Gurnah 2005: 78)

When Azad, one of the main characters of Abdulrazak Gurnah's novel *Desertion* (2005), makes this statement in order to console his wife in Mombasa and make her understand his frequent absences due to his merchant business, he addresses what he considers to be the central features of a Swahili trader's life. Indeed, in most of the writings on people from the East African coast, Swahili people are characterised as 'a seafaring and merchant people nurtured by contact' (Saleh 2002), and accordingly, much of the immense body of academic work concerning Swahili has concentrated in different ways on an examination of their translocal connections and resulting identities.

As Pouwels formulates it in his review of the book entitled *Les Swahili entre Afrique et Arabie* (Le Guennec-Coppens & Caplan 1991), the Swahili coast including Zanzibar is conceptualised as an "intersection of multiple influences and networks from which individuals derive their identities, and through which they establish and maintain relations with others in their complex social universe through various forms of exchange" (Pouwels 1991: 411). This conception indeed seems to lie at the heart of the majority of work on Swahili, which can broadly be divided in two strands. One big strand concentrates on Swahili origin and culture by especially tracing the historical relations across the Indian Ocean, with many arguments being based on archaeological findings along the east African coast (Chami & Msemwa 1997, Freeman-Grenville 1960). Another strand of work focuses on contemporary social relations and Swahili identity. In many empirical studies, marriage has served as an access to Swahili relations, and particularly the choice of marriage partners has been a central aspect in discussions on Swahili family relations and identity (cf. Le Cour Grandmaison 1989, Middleton 1992). Furthermore, Swahili literature and especially poetry have been taken as a medium through which to analyse the various sources of Swahili culture and identity (cf. Arnold 2002, Khamis 2000, 2004, Myers 1993, 2000, Vierke 2011). Finally, there are also a few recent publications on the Swahili diaspora, the majority of them focusing on their complex positioning as 'remigrants' in Oman and Yemen (cf. Al-Rasheed 2005, Valeri 2007, Walker 2008, 2011, Verne & Müller-Mahn 2012), but also debating the 'endurance' of Swahili culture and values in Europe or the USA (Saleh 2004, Topan 2006), the latter often strongly informed by the authors' personal experiences.

Overall, there is an understanding that the term 'Swahili' is generally used to refer to people from the East African coast who have a way of life characterised by the region's long-standing Afro-Arabic relations and the influences of the Indian Ocean, speak Kiswahili as their vernacular language and are Muslim. However, when talking to people who fit this categorisation, most of them would first consider themselves as a member of the city or region where they were born or have grown up, for example, *Waamu* (Swahili from Lamu), *Wamvita* (Swahili from Mombasa), *Wapemba* (Swahili from Pemba) or *Wazanzibari* (Swahili from Zanzibar's island Unguja). Whereas people in Mombasa more commonly use the term Swahili to refer to themselves, those who would be expected to do so in Zanzibar are far more hesitant as a result of the politics of 'swahilisation', part of the post-Ujamaa nationalism that made use of the term to strengthen a homogenous Tanzanian national identity. What soon becomes evident is that there are complex processes of inclusion and exclusion hidden behind the term 'Swahili', and that external and internal perceptions do not always match. Nevertheless, speaking of 'people like us', as Azad does in *Desertion* and as is indeed often heard for example in Zanzibar, Dubai or London, also hints at a certain Swahili identity, a sense of connection, and a shared understanding of what it is that holds these people together. As Ho points out in his impressive study on Hadrami mobility and genealogy across the Indian Ocean, 'such bonds exist and endure only so long as people continue to speak, sing, recite, read, write, and otherwise represent them' (Ho 2006: xxiii).

As well expressed in the quote, a central dimension and way of representing 'Swahiliness' is travel. Having played an important role in most of the intercontinental commerce between the coast of eastern Africa and the Persian and Arabian Gulf, the Indian subcontinent and Indonesia for well over a thousand years, Swahili can indeed be seen as travellers for millennia. However, there still exist a number of contesting views about the time of the first contact between the Arabian Peninsula and the East African coast. Some are convinced that constant and close ethnic, cultural, economic and political links were forged between the South Arabian kingdoms and the East African coast already during the early part of the first millennium B.C.; others rather rely on the first definite mention of Oman-Africa connections in 700-705 A.D. when rulers of the interior of Oman travelled to Africa to escape the Umayyad attacks (cf. Bhacker 1992: 25-27). What is certain is, that by 1700, after having successfully defeated the Portuguese in Mombasa in 1698, the Omani had managed to establish a loose hegemony over the Swahili coast. And, when Seyyid Said bin Sultan, the Sultan of Oman, in 1832 moved his capital to Zanzibar, this was followed by another wave of migration, especially of young Arab men, who started to make their living on the islands by engaging either in the plantation economy or in trade. Thus, smaller and bigger flows of migration, anticipated and followed by frequent movements back and forth between the East African coast, Southern Arabia or India, have long characterised life in the 'Swahili corridor' (Horton 1987).

During the second half of the twentieth century, it has been movements out of this 'corridor' that resulted in sizeable Swahili communities in Europe, the USA and Canada, as well as on the Arabian Peninsula. On the 12th of January 1964, only 33 days after Zanzibar was given full independence from the British, the new government was overthrown by a troop of mercenaries and followers of the opposition party, a political upheaval towards a socialist pro-African regime in which approximately 6000-10 000 residents were killed, and about 30 000 out of approximately 50 000 people of Arab origin were forcibly expelled or fled at their own initiative, many of them heading towards Arabia (Glassman 2011, Gilbert 2007). If the initial wave of Swahili migrants during this period was thus triggered directly by political persecution resulting from conflicts regarding Swahili identity, which led to this so-called Zanzibar Revolution, and was dominant in its aftermath, subsequent waves were more a product of economic problems and a sense of being extremely underprivileged. The often very close links these migrants are still maintaining to the East African coast as well as to Swahili people in other places are based on different forms of mobility, such as virtual communication and travelling, and therefore constantly (re)consitute and hold together what is now being called the 'transnational Swahili network' (Topan 1998, Horton & Middleton 2000).

Historically and even today, it is very difficult to examine travelling in the Swahili context without also looking at trade, which is both the content and context of the above quotation. According to Sheriff (1971: 10),

> 'trade is the pervasive theme in the history of the East African coast and is essentially a unifying economic and cultural force. It formed the basis of East Africa's foreign relations with the countries across the Indian Ocean, as depicted in the Periplus [of the Erythraen See] and by the Arab geographers. It gave rise to the Swahili city states which resembled the beads of a rosary, clustering in places and widely spread elsewhere, threaded together by coastal shipping'.

Referred to as a middlemen mercantile Muslim society, the Swahili's important position in the early commercial networks across the Indian Ocean as well as through the African mainland has been widely acknowledged (cf. Alpers 2009, Freeman-Grenville 1960, Horton & Middleton 2000, Middleton 2004). Especially during the time of Omani rule, Zanzibar was part of a network of ports held together by shared commercial interests: It was trade rather than a more centralised governing that formed the central occupation to hold together an Empire that consisted of interlinked city-states. In this respect, it has been shown that trade has not only long been regarded as a guarantor of wealth but has also enormously determined the hierarchisation of Swahili society by having strong effects on social status. As several authors have pointed out, trading connections have been extremely relevant to processes of identification throughout Swahili history (Cooper 1977, Kresse 2007, Le Guennec-Coppens 2002, Nurse & Spear 1985). This has been put most bluntly by Middleton, one of the most famous anthropologists working on the Swahili, when he states that, although 'the Swahili merchants view of their worlds never exactly mirrored the commercial and political actuality,[...] it gave reason and order to what they saw as their central position in world society as they knew it' (Middleton 2003: 519).

The abolition of slavery in Zanzibar in 1873 and the loss of control of territory in relation to the increasing colonial power of Germany and the British at the end of the 19th century weakened the Sultan's Empire and its economy. By the time when Zanzibar was officially declared a British protectorate in 1890, the Sultan of Zanzibar had already sold his mainland territory, a coastal stripe of ten miles that he had been granted by the Berlin Conference in addition to the islands. However, during the time of British rule, the Sultan remained a reigning but not ruling monarch and only little changes were made to the economic structure of the islands. It was following the Zanzibar Revolution, when the Revolutionary Government of Zanzibar abolished private enterprises and took control of all imports and retailings, that the previously very intensive trading connections came almost to a standstill. Only since 1984, when president Ali Hassan Mwinyi made efforts to restore Zanzibar's international relations, trade has become more liberalised again but it still suffers from high taxes and tariffs bound by the Tanzanian Revenue Authority (TRA). Thus, despite the relaxation of trade restrictions, its impact on the economy still seems very slight. Nevertheless, engaging in trade is still a dominant practice among Swahili people not only in Zanzibar but also along the East African coast more generally. Even though the strong economic decline as well as profound changes in the organisation of the Indian Ocean trade in the course of the last century, such as the mode of transport and the commodities, cannot be denied, as an ideology trade has remained its significance until today.

Contemporary Swahili trading connections are reviving old routes and further developing them through new translocal links to the diaspora. In this respect, it seems particularly fascinating to explore how long-standing trading routes and journeys have been adapted to contemporary political and economic conditions and how Swahili traders still make use of 'old' connections and networks in order to take advantage of the way the world economy works today. Although it is difficult to estimate their economic dimensions, it is evident that their imaginary and ideological basis, which is especially relevant to processes of identification, is still present. Talking to Swahili people today soon reveals that trading practices are still recognized as a worthwhile activity, not only for the anticipated profit, but primarily to see oneself as a part of a society that is based to a large extent on this very ideology. Nevertheless, despite a considerable agreement about the persistent relevance of trade in contemporary Swahili society, so far, there has only been very little engagement with the questions of how and why this is the case. By taking contemporary trading practices as the empirical focus of this study, this book therefore contributes to this often rather historically biased debate, providing a more detailed insight into the organisation, motivation and effects of trade in the Swahili context today.

Finally, the frequent absences due to his merchant business that Azad is trying to apologise for also mean frequent presences elsewhere, as well as frequent movements in-between. Being physically, virtually and metaphorically on the move is central to connect and constantly reconnect Swahili people in and between different places. Examining this mobility in its manifold dimensions therefore seems decisive in order to understand how ideas of 'Swahiliness' are negotiated, teasing out both possible tensions and unifying elements. Hence, an exploration of mobile practices and the ways in which Swahili people make their way in the world today not only contributes to the current themes in the field of Swahili studies, but also provides the detailed empirical insights from which to take further theoretical and conceptual ideas on the ways in which mobility impacts on senses of place, home, belonging and the creation of material and metaphorical spaces.

FOCUSING ON TRANSLOCALITY IN THE SWAHILI CONTEXT

Connecting these two strands - one of which being more theoretically informed and the other one driven by long-standing debates on Swahili culture and identity in Swahili studies - served as the main inspiration for the research underlying this book. First of all, this project can surely be understood both, as a response to the omnipresence of mobility in everyday life, and to the flourishing of mobility studies strongly visible in the social sciences and humanities in recent years. However, the recent burgeoning of mobility studies cannot be explained by the newness of mobility, although it might be right to state that its dimension and speed has increased in recent years - an assumption based particularly on a new attention to the role of mobility in contemporary Western societies. The Swahili case strongly 'contests assumptions and assertions that such highly connected geographies are recent developments' (Featherstone et al. 2007: 389), while at the same time illustrating the continuously high relevance of mobility for processes of identification today. Taking a look at Swahili history soon shows that mobility has been a major aspect of the construction of Swahili culture and identity for more than a thousand years, and is just as crucial today for the expansion of trading practices and the creation of relations between dispersed Swahili people. In connecting Swahili along the East African coast, the East African interior, the Arabian Peninsular, Europe and Northern America it thus offers an opportunity to tease out the meaning of mobility in a translocal setting that spans 'Western' and 'non-Western' contexts, clearly crossing the alleged binary between 'West' and 'rest'. Moreover, due to the high relevance of trade in this particular context, the empirical study includes human and non-human mobility, bringing to the fore the role of goods in the making and weaving together of different places, and thus opening the view to how the mobility of people, ideas and material objects relate to each other and come together in the constitution of translocal connections (cf. Crang & Ashmore 2009, Ogborn 2002). How do Swahili organise contemporary trading connections, what happens on the move, and how is this mobility of people, ideas and objects experienced and linked to feelings of relatedness and belonging? What is it that constitutes these translocal connections and what kind of spaces emerge out of them?

Although translocal connections have recently become one of the most popular topics in migration research, so far, the majority of research seems to concentrate on migrants' relations to nation states, their integration into the 'place of reception' or their social and economic effects on their so-called place of origin. In effect, despite much talk about links and connections, most empirical research is still conducted in 'nodes', limiting the examination of translocal connections to how people talk about them instead of how they *do* them. In this respect, bringing mobility to the fore and thus focusing the attention on what is actually happening on the move in-between places not only implies an important conceptual change, but also takes up the methodological challenges this shift necessarily involves (cf. Sheller & Urry 2006, Adey 2010a). How can we grasp mobility without loosing sight of the deep embeddedness of mobile practices in wider contexts?

Eventually, what came out of all this is a mobile ethnography of contemporary Swahili trading practices, as only this seemed to make it possible to live up to the central aim of this book: to ground the often overly abstract discourses of the relation between mobility, place and space in the lived experience of translocality in the Swahili context. Based on 'thick descriptions' of translocal Swahili connections a thorough understanding of the mobility of things, people and ideas, the ways in which they connect places and how this contributes to the creation of a specific translocal space can be gained, helping us to (re)conceptualise places, connections between places and the constructions of space under mobile conditions.

STRUCTURING THE TEXT: ARRANGEMENTS, MOVEMENTS, ENMESHMENTS

Despite the focus on connections and the resulting aim to emphasise the interconnectedness of different parts and alleged oppositions, to give some orientation to the reader it seems unavoidable to create a linear flow and divide the text in three main sections. The first part is named *Arrangements,* illustrating the development of the circumstances in which the whole study needs to be considered. Building on the ideas presented above, it therefore introduces the reader to the epistemological and methodological context of the study, making the construction and use of different ideas and views more comprehensible. Getting to grips with different understandings of translocality, the first chapter promotes a relational perspective, which shifts the focus from connected entities towards the connections themselves and the way they hold things and people in different places together. By doing so, it becomes possible to look at translocality as it emerges and as it effects complex mobile practices and processes of identification, instead of simply using it as a term to cover the existence of relations between (people in) different places. In order to avoid the structural and rationalist bend in most classical network images, the metaphor of the rhizome (Deleuze & Guattari 1976) is introduced as it opens up a way to better account for this relationality, and also for the complexity, multiplicity and heterogeneity of translocal connections that are constantly in the making.

Addressing questions of representativity and positionality, the second chapter attempts to create an understanding of the research practices and processes that have led to the production of this text, exemplifying what has been done, how it has been done and why. With trade playing a central role in translocal connections among Swahili people, contributing to its constitution as well as being one of its outcomes, trading practices are used as the empirical as well as representational access to the translocal Swahili connections. Drawing on anthropological discussions of ethnographic fieldwork, the geographical reception of this methodology, and the challenges posed to it by increasing mobility and interconnectedness, the book argues for a mobile ethnographic approach. This enables the researcher to get to grips with how translocal trading practices are actually lived and experienced, as well as how they influence senses of place and belonging. By accompanying traders and goods on the move, a complex and deep understanding can be gained of both the everyday business of Swahili traders and the flow of goods and ideas.

The second section is called *Movements* and consists of four empirical chapters each evolving along a different trading connection, referring to different types of mobility, different dimensions of trade, and pointing to different material and immaterial effects of the translocal practices (such as infrastructure, shops, availability of goods and negotiations of Swahili identity, senses of place and home). In this way, the diversity and multi-dimensionality of Swahili trading practices is expressed, whilst at the same time complex interconnections and interlacements are illustrated. Bringing together empirical experiences with theoretical debates, my intention is to give 'thick descriptions' of different mobilities, personalities, ambitions and places, as well as to contribute to the theoretical discussions that are addressed by them. An emerging theme, which is discussed from different angles in all of these chapters, is the role of economy and culture and the various ways in which they intermingle and intertwine on these different translocal connections, thus showing not only *how* trade is affected by 'culture', but also *how* Swahili culture is shaped by trade.

The third part is entitled *Enmeshments* as, according to the overall aim of this book – to both give a vivid portrayal of the way translocality is lived among Swahilis and to use this to contribute to theoretical reflections on translocality more widely –, here, empirical insights become enmeshed in the theoretical reflections of the first part in order to refine and enrich the understanding of translocality as a concept. While the first chapter elaborates how the different kinds of mobilities and cultural economic practices of material exchange lead to the construction of a translocal space, the main task of the final chapter is to develop arguments concerning the specific characteristics of 'living translocality'. These take up the most prominent recurring themes in the presentations of the four different connections, namely the role of location, the matter of distance and the negotiations of Swahili identity within the translocal connections under study. By drawing on empirical insights gained into these issues, the often overly abstract discussions of translocality and closely related ideas of transnationalism, relationality and cosmopolitanism are approached from an empirical perspective, confronting these concepts with the experience of those actually living translocal lives. This allows me to engage with some of the difficulties of recent contributions concerning the relation between mobility, place and space that often either miss the genuinely relational character of translocal spaces or overemphasise it by missing the continuing significance of distance and distinction. By giving a deep ethnographic insight into the effects of mobility on senses of place, home, belonging and the creations of material and metaphorical spaces, this book finally not only wishes to provide an empirically grounded contribution to theoretical reflections on translocality, relationality and respective conceptualisations of mobility and space but also attempts to provide a vivid insight into translocality – how it is lived, made meaningful and a way of being in the world – from a Swahili perspective.

ARRANGEMENTS

EPISTEMOLOGY:
A RELATIONAL APPROACH TO TRANSLOCALITY

At first sight, a research topic dealing with translocality in the Swahili context will probably be linked, by most readers, to the areas of migration studies, transnationalism or, more broadly, globalisation. Some might also wonder why I am hesitant to talk of a 'transnational Swahili network', a term generally quite excessively used in many areas of mobility studies (esp. concerning migration, economic relations and transport) (Dicken et al. 2001, Fowler 2006, Grabher 2006, Larsen et al. 2006, Portes & deWind 2007, Yeung 1998). So far, I have talked about the field of this research mainly as 'translocal Swahili connections'. In the following, I will start with this term, dealing with it more precisely in order to illustrate what I mean when using it. Despite the many different notions of the word 'translocal' which are currently used in academic writing on globalisation and migration and will be mentioned briefly, I will show how it can also be understood as a term indicating a relational and dynamic understanding of the world which emphasises movement and connection by both taking up central elements of the idea of mobility as presented above and corresponding to the impressions gained from the empirical material regarding the Swahili case.

Building on that, I show what this particular understanding of the term 'translocal' implies to the conceptualisation of connections and resulting networks by calling for an image that is less static and schematic than most conceptualisations of networks still dominant today. Building on ideas developed by Deleuze and Guattari (1976) and others, I discuss the differences of this perspective in comparison to more common approaches to networks, explaining why, more than thirty years after its publication, I still consider the rhizome as an enriching and very helpful metaphor for this current study on translocality in the Swahili context.

POSITIONING TRANSLOCALITY

'Translocality' is a term that increasingly occurs in articles, but is seldom elaborated in more detail. It seems almost taken for granted that the reader knows what is meant by it, although the use of the term as well as its context varies enormously. Whereas it is sometimes linked to a particular notion of the relationship between 'the local' and 'the global', it is also mentioned to refer to relationships between cities or simply to indicate that actions and practices somewhere are somehow connected to actions and practices elsewhere (cf. Grillo & Riccio 2004, Lachenmann 2008, Ma 2002, McKay 2006, Zhou & Tseng 2001). Moreover, the term has derived from different research fronts and is therefore understood differently in different disciplines. Nevertheless, translocality can be seen as a central concept illustrating the theoretical thoughts that underlie this study. In this section, by discussing its different uses, their strengths and weaknesses, I will thus come to introduce a particular understanding of the term which makes it a valuable lens through which to look at Swahili people and their connections between and across different places.

Between transnationalism and localism

Despite its popularity in migration studies, 'transnationalism' has first been used to refer to the widespread relations of international and non-governmental organisations (Vertovec 2009), pointing at the fact that they seem to operate on a level above and almost independent of nation states. In its common definition the term generically refers to 'occupations and activities that require regular and sustained social contacts over time across national borders for their implementation' (Portes et al. 1999: 219). Research interests in 'transnationalism' therefore include the border-crossing activities of non-governmental organisations, institutions and companies as well as migration.

Acknowledging that migrants increasingly live lives across national borders, the terms 'transmigration' and 'transmigrants' have emerged in migration studies emphasising the ties migrants forge and maintain with places in different nation states. Many scholars agree, that 'transnationalism represents a novel perspective, not a novel phenomenon' (Portes 2003: 874). It is Basch, Glick-Schiller and Szanton-Blanc who, in their book *Nations Unbound* (1994), first proposed to rethink notions of diaspora and regard 'transnationalism' as 'a process by which migrants, through their daily life activities [...] create social fields that cross national boundaries' (Basch et al. 1994: 22). Instead of seeing migration as a unidirectional movement, 'transmigrants' were in this respect considered as a special type of immigrant who develop and maintain multiple relations linking places of origin and settlement.

This idea has certainly led to a notion of disembeddedness of transnational activities, suggesting that living or acting transnationally means to be delocalised, free of attachment and emplacement. Some researchers soon criticised this, arguing that transnational migrants are not always in the air but must necessarily touch down somewhere (Ley 2004). Recently, academics have therefore become more interested in the relationship between the transnational and the local, as for some it is precisely in the localisation of more global flows that transnationalism is theoretically productive (Smith 2001). According to Guarnizo and Smith (1998), 'transnational practices, while connecting collectivities located in more than one national territory, are embodied in specific social relations established between specific people, situated in unequivocal localities, at historically determined times' (Guarnizo & Smith 1998: 11). In the same vein, Zhou and Tseng (2001), in their study on *Regrounding the 'Ungrounded Empires'* see localisation as the geographical catalyst for transnationalism. Following this argument, an increasing number of scholars now try 'to ground the discourse of the transnational in the place-making practices of the translocal' (Smith 2005: 243, cf. Brickell & Datta 2011: 3, Oakes & Schein 2006: 20, Smith & Eade 2008).

As Featherstone et al. (2007: 385) point out, stressing the importance of locality – the concept of trans-localities has become increasingly influential, capturing the ways in which trans-migrants are embedded in place, unable to escape their local context despite being 'transnational' (cf. Li 1998, Smith & Guarnizo 1998; Zhou & Tseng 2001). Translocal can here be seen as a 'form of local-to-local spatial dynamics' referring to the 'dynamic between localised lifeworlds in faraway sites' (Ma 2002: 133). Nevertheless, I would argue that, instead of really getting to grips with the relationship between what is called 'the local' and what is called 'the transnational', there is a tendency to simply name 'translocal' what has previously been called 'transnational' (cf. Vertovec 2009: 162). Whereas the latter has often been criticised for the remaining reference to the nation-state while stating that things happen beyond it, the former seems to address more openly the relations across different localities indifferent of its administration. Overall, much research in this area however still ends up looking at the influence of a locality of origin at a certain migrant community in their present location and often remains rather static and schematic (cf. Remus et al. 2008, Smith & Eade 2008).

Between 'the global' and 'the local'

As shown above, the term 'translocal' possesses a strong horizontal element when referring to relations across different locations. Moreover, also its vertical dimension can play an important role in respect to the debates on geographical scales (cf. Swyngedouw 1997, Brenner 2001, Marston et al. 2005). Over the last two decades, many have tried to grasp the relationship between 'the global' and 'the local', often assuming that these two can be regarded as separate entities (cf. Robertson 1995). Whereas 'the global' usually served as the globe-spanning universal, 'the local' has rather been seen as contingent and singular, responding to, struggling with or even resisting global processes (Cox 2005). Nevertheless, defining something as either global or local does hardly seem convincing (Dicken et al. 2001: 89). In his example of the railway Latour (1993: 117) illustrates very well how phenomena are both local and global and, as he points out, it is 'the intermediary arrangements between the global and the local that are the most interesting' (Latour 1993: 122). In this respect, there have been attempts to create a 'sense of the ways in which places are simultaneously made as both global and local without necessarily being wholly either' (Latham 2002: 116). Whereas Robertson introduced the term 'glocalisation' in order to connect the global to the local, others related the debate to ideas of hybridisation and creolisation (Pieterse 2003, Bhabha 1994, Hannerz 1996) analysing processes that were seen to be neither global nor local but bringing together or transcending different local phenomena. In a similar sense, the term 'translocal' now appears to refer to the connections between different 'locals' while not yet being considered as global (cf. Grillo & Riccio 2004: 99, Kraidy 2005: 155). Instead of consequently acknowledging the intermediary arrangements, fluidity and intermingling processes which will play a crucial role in this study and, by doing so, dismissing the search for a vertical ordering of the world, here, another step is added when somehow putting the 'translocal' in-between 'the global' and 'the local'. This idea therefore hardly manages to distance itself from the conviction that the world can be ordered according to clearly distinguishable scales that form a static framework or at least an explanatory device in which people – and their (attempted) movements – can be analysed (Inkpen et al. 2007, Dicken et al. 2001).

Between transgression and situatedness

Against the background of these two meanings of 'translocality', I now want to introduce a third understanding of the term which goes beyond the vertical and horizontal by focusing on bringing together movement and situatedness. Already when first used by Appadurai in 1995 the term 'translocality' addressed the ways 'in which ties of marriage, work, business and leisure weave together various circulating populations with kinds of locals' (Appadurai 1995: 203). Instead of simply referring to a single place or a collectivity of places, translocality can also be understood to correspond to a rather abstract space existing of the manyfold connections within it (Mandaville 2001). Following Mandaville (2001: 6), 'translocality is primarily about the ways people flow *through* space rather than about how they exist *in* space'. Having become popular by Castells idea of 'spaces of flows' instead of 'spaces of places' (Castells 1996), a strong emphasis is put on places as unbound and crosscut by flows of people, things and information (Allen et al. 1998, Amin & Thrift 2002: 3, Amin 2002). Most prominently in geography, Massey has argued that it is the connections and relations between extensive flows of people, capital and power that constitute places, and instead of understanding them as internally bound and separated from the outside, she promotes to account for places as consisting of both relations 'within' the place and the manifold connections reaching far beyond it (Massey 2004b: 6). To view a place as a translocality therefore means to understand the (re)construction of places through the movements of people, material objects and ideas through places (cf. Amin 2004, Appadurai 1996, Hannerz 2003, Smith 2001).

Nevertheless, as Mandaville points out, 'the notion of locality is included within the term in order to suggest a situatedness, but a situatedness which is never static' (Mandaville 2001: 50). In this respect, translocality is neither reduced to mobility nor to any fixed outcome, but it incorporates the tension and interplay between mobility and situatedness, movement and stability. In a recently published edited volume *Translocality – The Study of Globalising Processes from a Southern Perspective,* Freitag and von Oppen (2010a) further elaborate this understanding of translocality. As an object of enquiry, translocality is seen to 'designate the outcome of concrete movements of people, goods, ideas and symbols' (Freitag & von Oppen 2010b: 5). As a perspective, translocality indicates a relational and dynamic understanding of the world, by focusing on connections and mobility and thus looking at place and space from the perspective of movements, 'highlighting the fact that the interactions and connections between places, institutions, actors and concepts have far more diverse, and often contradictory effects than is commonly assumed' (Freitag & von Oppen 2010b: 5).

In effect, places are more and more understood as 'meeting places' (Allen et al. 1999: 33), 'places-of-interaction' (Amin et al. 2000: 8) and interminglings of translocal connections. The idea of translocality is here to research places as 'switchers', middles and intermediaries in networks (R. Smith 2003b: 576) rather than 'centres of command and control' (Sassen 1991). In this respect, Smith has developed a particular notion of 'translocalities', emphasising that what is considered as local is not detached from other places but 'a fluid cross-border space in which social actors interact with local and extra-local institutions and social processes in the formation of power, meaning, and identities' (Smith 2001: 174). 'Places are about relationships' rather than simply containing things, and are seen as 'travelling, slow or fast, greater or shorter distances, within networks of human and nonhuman agents' (Sheller & Urry 2006: 209).

Instead of trying to locate trading connections somewhere between 'the global' and 'the local' or aiming at determining how they link locations regarded as separate entities, this reconceptionalisation of place leads to an idea of translocality, which no longer simply emphasises the links between places but positions places as products of diverse forms of connection and relations aiming to explore their links as their constitutive elements. Therefore, this idea of translocality incorporates both mobility and emplacement and is about studying what flows through places as well as what is in them, by explicitly addressing its relationality and linking both dimensions through multi-directional processes of connecting, cutting, layering and enmeshing. By proposing a more open and less linear view on the manifold translocal connections, translocality considers the transgressions of boundaries between spaces of very different scales and types, thus clearly distancing itself from the close reference to nation-states as indicated in the term 'transnational' (Freitag & von Oppen 2010b: 6).

As much as translocality is seen as a phenomenon that can be described, it also refers to a particular condition, a particular way of being in the world, which is characterised by the tension and interplay of mobility and situatedness. In order to avoid an overemphasis on movement, flows and transgression, this perspective therefore also entails a strong interest in the ways in which translocality is dealt with and experienced by people actually living in this condition (cf. Freitag & von Oppen 2010b: 8). Hence, following this understanding of translocality opens up a way to look at the ways in which the movement and manifold connections of people, material objects and ideas within Swahili trading practices are actually lived. Moreover, applying this understanding of translocality to this study of contemporary Swahili trading connections seems fruitful seeing them as diverse links and paths through different parts of the world which cross each other, meet and mesh and, by doing so, play an enormous role in the constitution of particular places and spaces (cf. Featherstone et al. 2007: 383–384).

SHORTCOMINGS OF THE 'NETWORK'

Thinking of the term 'translocal' as entailing a strong sense of the relevance of relationality and connectivity focusing on mobility and situatedness and the manifold relations between people, objects, ideas and places, an association with networks is not surprising. Nevertheless, 'network' has become a very ubiquitous term that seems omnipresent in much of the research on globalisation, transnationalism and migration so that it has to be made clear what kind of image of the network one is following when using the term according to a translocal perspective.

Regarding social interactions as core elements of society, it was Simmel who already in his thoughts on sociology and the construction of social forms (Simmel 1908, see also Simmel 1890) developed the key idea of network concepts. Since then, the research on social networks has developed into one of the big multidisciplinary research fields in the social sciences, based to a large extent on the methodic repertoire of social network analysis, substantially advanced and refined especially in the 1970es (cf. Stegbauer & Häußling 2010). In geography, it was especially the influential works by Peter Hagett (Haggett 1965; Haggett & Chorley 1969) that made it a popular concept for the positivist human geography during the 60s and 70s. From the early 90s onwards, the interest in networks has grown even stronger than before, influenced by the works of Granovetter (1973, 1983, 1985) and extremely pushed by writings such as *The Network Society* by Castells (1996) and the assumption that globalisation will above all lead to an extension, deepening, speeding up and increased impact of global networks. As a result of this fashioning of networks, it has turned into an ambiguous term, and many have started to mourn about its 'conceptual elasticity' (Grabher 2006: 164) seeing it as 'more of a chaotic conception' which is much-abused (Yeung 1994: 475). Nevertheless, it remains an evocative metaphor and seems unavoidable when talking about relations and connections between different people, objects and places. Therefore, instead of trying to find a universal definition, I will discuss some of the core elements of the prevalent images that can be identified in recent writings on networks (see e.g. Bebbington & Kothari 2006, Emirbayer & Goodwin 1994, Fowler 2006, Grabher 2006, Hughes 2007, Leitner et al. 2002a, 2002b, Nicholls 2009). By also dealing with their criticisms, it soon becomes clear that following a translocal understanding of Swahili connections makes it necessary to move beyond these network images, formulating demands that need to be fulfilled by an image which aims at reflecting and corresponding to the translocal perspective as developed above.

In the literature discussing social network approaches, it is striking that networks are generally regarded as a bounded formation consisting of a limited number of actors who are related to each other and use these relations for the exchange of information or goods, so that their analysis concentrates on the number of actors, the function of the network as well as its structure (see e.g. Kilduff & Tsai 2003, Scott 2000, Wassermann & Faust 1994). This image of the network is closely connected to the seminal works of Granovetter (1973, 1983, 1985) who, as a response to the economic debates on transaction costs, developed his idea of embeddedness and the differentiation in strong and weak ties and, by doing so, formed what has become known as the new economic sociology. Focusing on individual actors and their interpersonal connections he emphasised the role of personal relationships (Swedberg & Granovetter 2001) leading to micro-level analysis of economic organisations. On a broader scale, it is probably Castells' understanding of networks as advanced in his trilogy (1996, 1997, 1998) that has become most influential in evoking a view of the world as a network. Studies on global flows and networks therefore often go together (Bebbington & Kothari 2006), and especially in works on the global economy and migration the term network dominates to indicate far-stretched connections. As Hughes points out, 'many studies [on knowledge circulation] adopt the metaphor of the network to capture the movement of knowledge through the global economy' (Hughes 2007: 531). Taylor, for example, speaks of the 'world city network' (2004) analysing how cities are related to each other by providing a quantitative insight into the connections between major world cities.

Although Taylor as well as similar studies have so far concentrated on big cities playing an important role in the global financial economy and hardly consider the connections between more mundane and inconspicuous cities, introducing the idea of city networks has nevertheless led to a new way of thinking about cities more generally – foregrounding exchange and circulation in contrast to ideas of the city as a bounded entity (cf. Amin & Thrift 2002, Beaverstock et al. 2000, Taylor et al. 2001). Furthermore, this has however also led to the idea of networks as an organisational form, which can be analysed using network governance approaches. Overall, it can therefore be observed, that whereas one strand of network analysis is still very much concerned with the social relations between individual actors and their effects (social network approaches), networks also serve as a strong image to conceptualise flows and movement of all kinds (network governance approaches and, more generally, what could be called a network-view on the world). Especially concerning the latter, a detailed description of how these kinds of networks look like is often missing.

As Emirbayer and Goodwin have remarked, the dominant images of networks are generally characterised by a 'structural bent' (Emirbayer & Goodwin 1994, see also Bair 2008) and, according to Grabher, 'social network as well as governance approaches [offer] models for conceiving (or at least implicitly assuming) the fabrics of socio-economic life that could be assorted neatly onto different scalar levels' (Grabher 2006: 179). Almost appearing to be possessed by the aim of classifying networks, strongly relying on causality and functionality, there is still only very little attention paid to the explanation of how networks and relations come into being and how this affects the different actors involved. Especially social network analysis often examines the morphological and interactional characteristics of networks by using increasingly complex quantifications, matrix representations and line-graphs to map and portray social relationships as a network structure (cf. Kilduff & Tsai 2003, Scott 2000, Wassermann & Faust 1994). Relying on sociometrics and software tools to rigidly analyse characteristics of social relationships, the human being often disappears in these calculations, turning into a static description detached from the social context and any interest in the role of cultural norms and discursive understandings (Grabher 2006: 174, Häußling 2009, Stoller 2003). Even in case studies that follow a more qualitative approach, networks are often imagined as given, so that their analysis mainly attempts to determine whether and how an individual's position within the network effects the degree of influence that can be exerted within it (Leitner et al. 2002b: 283). Examining the centrality of actors, it is assumed that this can be determined by looking at the number and frequency of relations whereas their qualitative nature is hardly been considered. As Stoller (2003: 199) states, 'although this literature is inherently interesting, especially for [researchers] with formal training in mathematics, it is of only partial relevance to a study of the dynamics of [in his case] New York's West African trading networks'. Despite the aim to 'transcend the atomistic description' (Dicken et al. 2001: 91) by looking at networks, a big part of the job still consists of identifying the actors, their relations and their structural outcomes. And, the predominant assumption underlying this work still appears to be that it is previously autonomous actors that form relations, nodes creating ties, instead of following a relational perspective in which relations have a big impact on the creation of actors as well. Also more recent images of networks as 'overlapping and contested material, cultural and political flows and circuits that bind different places together through different relations of power' (Featherstone et al. 2007: 385) fail to fully acknowledge the transformative character of relations as taken up in a genuinely translocal perspective when talking of 'binding' and not 'creating'.

Moreover, especially in migration research, networks are generally regarded as a facilitator of migration. As early as 1989, Boyd (1989: 641) already remarked that 'networks link populations in origin and receiving countries and ensure that movements are not necessarily limited in time, unidirectional or permanent'. Until today, in this context, networks are mostly seen as something positive and supportive, giving the migration process a certain structure and order (cf. Chelpi den Hamer & Mazzucato 2010, Portes & DeWind 2007, Vertovec 2009). On the other hand, disorder, insecurity, and situativity are hardly being accounted for in these classical network images.

After all, these common conceptualisations and representations of networks do not seem very satisfying in the light of the theoretical reflections on translocality. An image of networks which successfully wants to take up the elements formulated in the understanding of translocal as developed above, first of all, would need to depart from the static view of 'transport without deformation' (Latour 1999: 15) and foreground the transformative and processual dimension of translocal connections. Expressing a relational understanding, it has to, as Grabher illustrates, 'stretch, crumple and blur the familiar tie-and-node imagery of networks' (Grabher 2006: 167) and contrast the rather clear-cut view on network formations in the dominant network approaches. This means that all actions and possibilities would have to be seen to derive out of and depend on connections instead of being inherent characteristics of discrete elements (cf. Mol & Law

1994, Hetherington 1997, Law & Hassard 1999). Instead of following a generic sociometric conception of networks, it needs to account for 'the multidimensionality of network rationalities and the multiplicity and fluidity of network relations' (Grabher 2006: 178). In a similar vein, a translocal network needs to be more open to what forms the connections by incorporating the flows of people, things and ideas. Seeing translocal practices as the product of manifold connections and intermediary arrangements between the human and non-human world, it therefore has to shift beyond established dualisms such as structure/agency, subject/object or human/non-human (Hinchliffe et al. 2005: 651). And finally, another important aspect is to not only proclaim it, but consequently transcend scales by challenging the perspective that assumes given and clearly distinguishable scales and, depending on the topic, promotes either one or the other as more important. Instead of putting the world in layers, categories and systems, a translocal image of the network has to reflect its complex, dynamic and promiscuous character that cannot be captured by those notions (Latham 2002: 138). This will then lead to an image which can offer more plural accounts of what is involved in the construction of translocal processes between movement and situatedness, recognising that connections are not givens but always actively made.

THE METAPHOR OF THE RHIZOME

When understanding 'translocality' as a perspective which leads to an image consisting of manifold and heterogeneous connections between places, interwoven paths and crossing flows, this idea fits very well to the metaphor of *The Rhizome* as developed by Deleuze and Guattari (1976). Written as the introduction to *Mille Plateaux* (1980), they suggest the rhizome as a toolkit that helps to visualise processes of networked, relational, and transversal thought (cf. Colman 2005: 231). They oppose the rhizome to the vertical structure of the arborescent model that, in their view, stands for rationalistic, dualistic and hierarchical/scalar thinking, and is thus seen to evoke everything that rhizomatic thought attempts to turn away from. In my opinion, the metaphor of the rhizome serves as a promising starting point from which to look at dimensions of translocal Swahili connections, which, so far, have not gained enough attention, or, at least, have not been addressed in this particular way. In particular, it enables me to account for the complex, dynamic and genuine relational character of translocal connections, getting to grips with the questions of how the connections are made, what they consist of and how they manage to endure and seem like a translocal space.

Thinking of the world as a meshwork of multiple and branching roots deriving from multiple sources that become entangled and intermingle with one another, Latour considers rhizome to be the 'perfect word for network' (Latour 1999, see also Lynch 1995: 169). While others question if 'rhizome' really evokes the best image – Ingold for example prefers the image of the fungal mycelium (Ingold 2006: 13, Ingold 2007: 41) – I will restrict the following elaborations to the characteristics that have been ascribed to the metaphor of the rhizome instead of pondering on the best term to capture them.

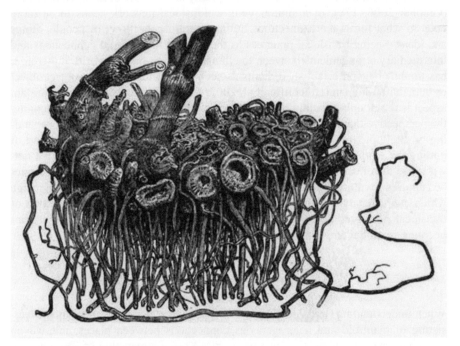

Fig. 1: Rhizome of Cimicifuga racemosa (Source: Felter & Lloyd 1898, Fig. 72)

A very evocative 'rhizomatic' image is developed by Serres (1991), who introduces a network consisting of points as intersections of relations, as their beginning and ending. Here, lines are understood as connections between two or more points, as results of the relations between the points. Continuous changes of position depending on the contents and power of the points and relations lead to its ever-changing and fluid form holding together as well as being held together by a multiplicity and complexity of lines. In this image it is not necessarily the shortest line being the easiest and best to get from one point to the next. Therefore, sometimes even by chance, a possible path is chosen from other possible paths (Serres 1991: 11–14). According to Serres, this makes it possible to conceptualise the world by acknowledging its complexity that derives out of the multiplicity of thinkable and doable possibilities. Complexity, thus, is not a hindrance to the acquisition of knowledge, no solely descriptive judgement, but develops into an excellent device of knowledge and experience (Serres 1991: 23). So, while it is unavoidable for theories and abstractions to homogenise and simplify the world, the rhizome at least attempts to provide a way to talk about, appreciate, and practice its complexity (Law 1999: 8–10). This also manifests itself in the language in which *The Rhizome* is written by Deleuze and Guatarri and others who further develop the image. Although its poetic style sometimes makes the elaborations seem chaotic, arbitrary and unfamiliar, this can be comprehended in respect to this central concern: trying to do justice to complexity and grasp the best we can what is actually happening by ‚appreciating the multidimensionality of network rationalities and the multiplicity and fluidity of network relations' (Grabher 2006: 78).

The metaphor of the rhizome is not new, but it is especially in the course of the last decade that it has gained considerable influence and is now increasingly mentioned in geographical discussions on networks, though still not being as common as the tie-and-node imagery (cf. Doel 1999, Grabher 2006, Hess 2004, Jones 2008, Powell 2007, Schlottmann 2008, Smith 2003a). Moreover, the image of the rhizome appears in much of the more recent work published under the heading of Actor-Network-Theory (cf. Latour 1996, Mol 2002) where it has been guiding the empirical research of science and technology studies and also served to regard phenomena such as the market or anaemia (Callon 1999, Mol & Law 1994) to mention only the most famous works in this respect. Despite being a lot less visible in research on translocal migration and mobility, I argue that it could also prompt empirical research that is better able to address the complexity, heterogeneity and mobility in translocal connections than studies on migrant networks in the more classical sense. Looking at some of the central characteristics of the rhizome as outlined by Deleuze and Guattari almost thirty-five years ago, I will therefore point out what I consider as its main advantages in comparison to the more common tie-and-node image of networks, in order to illustrate how it can still serve as an enriching metaphor to conceptualise translocal Swahili connections.

Relationality

The first advantage of the metaphor of the rhizome is that it is based on a genuine relational understanding of the world. Although there have been numerous calls for a relational thinking in geography since the mid-1990es (Allen et al. 1998, Bathelt & Glueckler 2003, Massey 1994b, 2004a, Graham and Healey 1999), it still seems to be a lot easier to put relationality on the agenda than to actually pursue it in research practice (Massey 2004a: 3). As Earman already noted in 1989, 'there are almost as many versions of relationism as there are relationists' (1989: 12). In my view, two main strands of relational thinking can be distinguished. The first strand is based on the widespread assumption that relationality presupposes binaries and that relational thinking thus needs to attend to the importance of interconnections between the different entities. As Yeung states, 'thinking about relationality necessitates an analytical movement away from abstract phenomena to examine the interconnections between discrete phenomena and to transcend their dichotomisation' (Yeung 2005: 44). In effect, it can be observed that most of this work is relational only in a thematic sense, meaning that it is concerned with the relations among actors and structures or different institutions and the ways in which they are intertwined and interact at different scales without understanding actors und institutions as the result of relations (cf. Amin 1998, Dicken & Malmberg 2001, Bradbury & Lichtenstein 2000: 551, Staeheli 2000: 9). This argument is also pursued by Emirbayer in his manifesto for a relational sociology, when he states that interaction is often mistaken as a relational point of view (Emirbayer 1997: 285). In contrast to that, he suggests following the transactional perspective, which acknowledges that the very terms or units involved in a transaction derive their meaning, significance and identity from the (changing) roles they play within that transaction. The important step in this second strand of relational thinking is to take the transaction as the primary unit of analysis rather than the constituent elements themselves. As Cassirer (1953: 36) put it:

> '[Things] are not assumed as independent existences present anterior to any relation, but [...] gain their whole being [...] first in and with the relations which are predicated of them. Such 'things' are terms of relations, and as such can never be 'given' in isolation but only in ideal community with each other.'

Thinking relationally in this second sense requires to reject the idea that one can posit discrete, pregiven units and, instead, to see 'relations between terms or units as pre-eminently dynamic, unfolding and ongoing processes rather than as static ties among inert substances' (Emirbayer 1997: 289, see also Hart 2004: 98). As Berndt and Boeckler argue in their recent article on the adaptation of relational thinking to economic geography,

> 'such a reading of the world shifts our attention from elements and identities towards relations, from traits to processes. Society then no longer appears as a conglomerate of (self)identical elements, [...], but rather as relational assemblages without origin, whose constituent parts can only be conceptualised as effects of contested processes of differentiation realised practically and loaded with power.' (Berndt & Boeckler 2008: 3)

Consequently, as Rajchman argues 'we should no longer think in terms of lines going from one fixed point to another, but, on the contrary, must think of points as lying at the intersection of many entangled lines [...]' (Rajchman 2000: 100). This understanding, then, leads to what Ingold illustrates as a 'field not of interconnected points but of interwoven lines, not a network but a *meshwork*' (Ingold 2006: 13).

While the interactional notion of relationality is still dominant in most network approaches, it is this second strand of relational thinking that is inherent to the metaphor of the rhizome and corresponds to what Deleuze has in mind when he emphasises:

> 'I tend to think of things as sets of lines to be unravelled but also to be made to intersect. I don't like points.... Lines aren't things running between two points; points are where several lines intersect. Lines never run uniformly, and points are nothing but inflections of lines. More generally, it's not beginnings and ends that count, but middles.' (Deleuze 1995: 160–61)

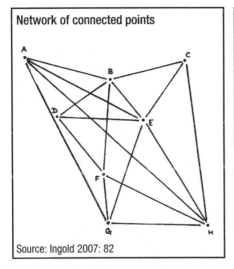 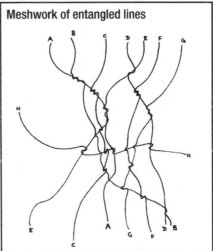

Fig.2: Illustration of relational thinking

Heterogeneity

This dislike of beginnings and ends that is expressed in the quotation above can more generally be understood as a general dislike of oppositions and binary thinking. Instead of seeing things as either one or the other, with the image of the rhizome Deleuze and Guattari attempt to overcome all kinds of dualisms by finding a way to address the ‚middles' (cf. Deleuze 1995: 161) This so-called 'associationalist thinking' (Murdoch 1997: 321) is the second advantage which I wish to elaborate here, as it enables us to see translocal connections as heterogeneous and multidimensional associations in which attention needs to be paid to all the elements in order to understand the way they are held together.

As Sayer (1991) has argued in his article on *Deconstructing Geography's Dualisms*, dualistic thinking has to be considered as problematic because it polarises whole fields of concepts by dividing theoretical perspectives into two distinct and incommensurable parts, thus leading to a fractured and very simplified view of the world. Nevertheless, it is still dominant in most socio-spatial analysis. From its non-dualistic standpoint, the 'rhizome' takes up the aim of overcoming the established binary juxtapositions of structure/agency, global/local, human/nonhuman, economy/culture by 'focusing on how things are "stitched together" across divisions and distinctions' (Murdoch 1997: 322). Understanding entities, their form and attributes, as results of their relations with other entities (Law 1999: 3), as Murdoch illustrates in his article on *A Geography of Heterogeneous Associations* (1997), the 'rhizome' allows us to follow 'a route which permits a careful negotiation of the extremes and the development of a more sophisticated "in-betweenness", a more nuanced, not quite here or there "kind of approach"' (Murdoch 1997: 321). This idea is also taken up by Latour and Callon who both see a central purpose of the metaphor of the rhizome in helping us apprehend the way in which people and things work together to create the world and therefore, to analyse the nonhuman in the same register that we examine the human (Callon 1986: 200, Latour 1987, 1999). This call for a symmetrical engagement with humans and non-humans entails 'to assert that *everything* deserves explanation, and more particularly, that everything that you seek to explain or describe should be approached in the same way' (Law 1994: 9).

Especially in discussions on Actor-Network Theory, this idea of symmetry has caused much controversy and has been accused of leading to a dangerous relativism (Laurier & Philo 1999: 1055). However, treating non-humans symmetrically in methodological terms does not necessarily imply to also assume their ontological symmetry (cf. Mutch 2002: 489). Even if it can be argued that, in the end, it often might be human forces driving heterogeneous sets of relations, this does not need to have the effect of assuming this *a priori* and denying the combination of human and material resources in networks. In this respect, as Whatmore states, a 'rhizomatic' view simply 'extends the register of what it means to generate materials from one in which only talk counts, to one in which bodies, technologies and codes all come into play' (Whatmore 2003: 97). This indeed seems crucial when trying to understand translocal trading connections in which both humans and material objects play equally decisive roles.

Overall, incorporating a symmetrical approach seems to enable a more nuanced and differentiated understanding on how translocal networks are constantly built and rebuilt through processes involving both humans and material objects. Instead of sticking to the extreme divisions into human/non-human, social/material, subject/object, by forcing us to look more closely at the associations which give rise to these concepts, a ‚rhizomatic' perspective helps us to see these divisions as outcomes and effects of processes of category making (cf. Murdoch 1997: 334). Accordingly, it should encourage us to see translocal trading connections 'as the product of unruly materially heterogeneous assemblages' (Hinchliffe et al. 2005: 651) challenging us to make sense of the ways in which various actors and elements come together by being particularly attentive to what will be linked to what and how these linkages will be forged.

Processuality

Any attempt to visualise translocal connections can only result in the representation of a momentary state of being of an actually fluctuating situation (Serres 1991: 9). This processual nature of connections is taken even further in the idea of the 'rhizome' as it is meant as an image of thought which thinks of the world as a meshwork of multiple and branching roots 'with no central axis, no unified point of origin, and no given direction of growth (Thrift 2000a: 716). Accounting for the partial and constantly changing nature of these meshworks of connections, expressed by Deleuze and Guattari in 'lines of flow and flight' (Deleuze & Guattari 1977: 34), and seeing the 'rhizome' as a series of transformations and translations (Latour 2005: 108) is the third aspect that deserves a closer attention.

Contrary to a static conception of 'networks' at a given point in time, which often goes hand in hand with the assumption that this is how a network actually is and, therefore, can be analysed as such, the 'rhizome' is imagined as a continuously changing process, thus reflecting more accurately the ways in which translocal connections are developed, changed, expanded and restricted simultaneously in different parts. This multiplicity, which is advanced by Deleuze and Guattari as a central principle of the 'rhizome', corresponds to its constant proliferation as expected and unexpected connections are made, leading to multiple shapes and functions (Deleuze & Guattari 1977: 21, Henaff 2005: 184). The rhizome has no linear, bounded or fixed character, and is meant as an 'anti-genealogy' which exclusively consists of lines which continually change, expand, resume, mutate and evolve into new constellations of connections, forming what Serres and Latour call a 'syrrhese' (Serres & Latour 1995). Furthermore, addressing the fluidity and vitality of connections, any connection of a rhizome when broken or destroyed, in contrast to leading to a significant rupture in the network, is first of all seen to bring about new possibilities of different connections, growing in different directions or connecting and modifying other parts (Deleuze & Guattari 1977: 16). What becomes clear by looking at this idea of multiplicity is that a 'rhizomatic' language generally emphasises the becoming over the being by promoting the processual character of embeddedness and social relations (Ingold 2006: 15, Hess 2004: 179).

Nevertheless, this focus does not imply that the elements are flowing infinitely within connections. Instead, there are lines of segmentation that stratify, mark, assign, organise and territorialise the 'rhizome' (Deleuze & Guattari 1977: 16). Movement and situatedness induced in the understanding of translocality are therefore brought together by indicating processes of embedding, through growth and the amalgamation of heterogeneous elements, as well as processes of disembedding. As Inkpen et al. emphasise, 'entities of which networks are composed require the continual flow of processes to stabilise their nature' (Inkpen et al. 2007: 3). Similarly, Ingold argues, that it is 'the dynamic, transformative potential of the entire field of relations within which beings of all kinds, [...], continually and reciprocally bring one another into existence' (Ingold 2006: 10).

Understanding translocal connections 'as a verb, a process' (Law 1992: 2), the metaphor of the 'rhizome', therefore allows us to examine 'how it becomes' instead of 'how it is' by being especially attentive to its development and changes in time and space. Despite presumingly analysing how established networks work, this perspective thus opens up more productive ways of dealing with the questions of how and why networks are created and maintained, and by doing so, broadens the view to the 'significant amounts of energy, resources and organisation that go into sustaining translocal lives' (Conradson & Latham 2005: 228).

Connectivity

All of the three dimensions that have been addressed so far – the relational, heterogeneous and processual character of the image of the rhizome – have emphasised the crucial role of connections and especially their making in order to understand translocal phenomena. For this reason, the forth advantage of a rhizomatic perspective which I want to stress here lies in its strong empirical focus on the connections themselves.

Concentrating on the 'middles' enables us to account for the heterogeneous associations of which relations are made, evoking an understanding of how and why particular people, things and ideas come together in the making of translocal connections. Furthermore, attention needs to be paid to the ways in which these relations are formed, reformed, redirected and changed in order to grasp the fluidity of networks as they are always in the process of becoming. Finally, according to the idea that 'points are where several lines intersect' (Deleuze 1995: 160), a focus on the ways in which different connections entangle and intertwine is needed in order to illuminate how places are created and constantly recreated.

Following Deleuze and Guattari, the 'rhizome' is based on a general connectivity of the various lines, so that any part of a 'rhizome' could possibly be connected to any other (Deleuze & Guattari 1977: 11). Whereas by far not all of these possible connections are actually anticipated, and while there exist different degrees and qualities of connections (cf. Law 2006 [1997]), what remains is an overall openness towards their formation. Understanding the 'rhizome' as being made up of multidimensional and constantly evolving connections that cannot be captured in any kind of static framework, it possesses neither a centralised nor a hierarchic structure that could be assumed prior to any research. As Henaff illustrates, with every site being or at least possibly being connected to all other sites, every point of the 'rhizome' becomes a centre in the multiple intersections of the meshwork, so that 'each [...] point implies the [...] network and the [...] network is nothing without the multiplicity of the individual sites' (Henaff 2005: 181). Therefore, the task of the researcher becomes to unravel the connections and uncover how translocal connections are built and maintained as well as showing the effort and power that is needed to do so by following translocal connections all the way along the different lines.

This approach has been criticised for not paying enough attention to structural power relations and, in effect, becoming a 'mechanistic framework that atomises agents and focuses solely on the links between them' without a sense of wider social processes and global economic forces that constitute these relationships' (Dicken et al. 2001: 105, Leitner et al. 2002b: 285). This criticism however can only be justified from a standpoint that presupposes the existence of clear power structures and thereby reduces any micro-level analysis to the effects of generalisable structural processes. On the contrary, 'rhizomatic' research tries to remain a general openness rejecting the view that the field of study is prearranged into certain levels of which some determine what goes on in others. As Law points out, 'it will not distinguish before it starts between those that drive and those that are driven' (Law 1994: 13). Instead, it attempts to gain detailed stories of associations and interdependencies, which then allow for an interpretation of the hierarchies and power relations as they actually occur in and characterise the connections (Latour 1986: 276).

On the one hand, connections have to be dismantled in order to reveal their composition, unequal power relations, ruptures, and constraints within them. On the other hand, it is also of interest how such contested meshworks are held together in a way that makes them appear as a unity (Kracauer 1995: 251–252). In this respect, there is a central concern to:

> '[understand the ways in which] actors mobilise, juxtapose and hold together the bits and pieces out of which they are composed [...] and how they manage, as a result, to conceal for a time the process of translation itself and so turn a network from a heterogeneous set of bits and pieces each with its own inclinations, into something that passes as a punctualised actor' (Law 1992: 6).

Although all translocal networks are the effect of multiple heterogeneous connections, these complexities are not always laid open. Therefore, the detailed research on connections that is implied when looking through a 'rhizomatic' lens also helps to create an understanding of how the network is performed and made meaningful in processes of punctualisation. Dealing with the appearance of unity on the basis of a strictly inductive and open research process finally allows for some empirically informed generalisations about the characteristics of translocality. As argued in the beginning of this introduction to the metaphor of the 'rhizome', this again expresses the attempt to bring as much of the empirical complexity as possible, to bring more of the meshwork itself, into the more general discussions and abstractions, which can then be fed into wider theoretical and thematic debates on the specifics of living in a translocal space.

1ST HALT: TRANSLOCALITY REARRANGED

It has been the aim of this chapter to introduce the epistemological context of this study by creating a relational and dynamic understanding of 'translocal Swahili connections'. Going beyond seeing the 'translocal' either as an additional scale between the global and the local or as an indicator of local-to-local relations, I have suggested a more relational understanding of the term, which entails a reconceptualisation of places as effects of connections and foregrounds the tension and interplay between transgression and situatedness inherent in translocal connections. As a perspective, translocality seems fruitful as it turns the attention to translocal Swahili connections as diverse links and paths through different parts of the world which cross each other, meet and mesh and, by doing so, play an enormous role in the constitution of particular places and spaces.

Though connections are usually discussed in the context of networks, I have shown how such a relational approach to translocality cannot be visualised in the form of classical tie-and-node images in which nodes remain to be conceived as discrete entities. Thus, I have highlighted how the metaphor of the 'rhizome' is able to contribute to a more appropriate understanding of translocal connections than the ones entailed in more structural network approaches, by recognising and directly addressing the relational, dynamic, and promiscuous character of translocal connections. This understanding will be crucial in order to go beyond an analysis that focuses on the interactions between separate entities in a clearly defined, rather static and layered framework, and instead manages to imagine translocality as a living organism characterised by manifold and heterogeneous translocal connections which are constantly evolving and changing, merge and mingle, not fitting into any clear pattern or grid. Only a 'rhizomatic' approach is able to account for a context in which it is common to meet Swahili who, for example, were born in Zanzibar, live in London, spend their holidays in East Africa – where they are constantly told to be Omani Arabs – have relatives spread over many different places, most of them regularly being on the move meeting and exchanging goods, so that it is not possible anymore to give these connections a clear shape with a designated origin, beginning and end.

Before empirically getting to grips with these connections, their making and how they succeed in holding Swahili together, in the following, I will first discuss in more detail how these epistemological thoughts can also be reflected in the methodology. Against the background of this particular study on the lived experience of translocality in the Swahili context, the next chapter will thus illustrate the methodological implications and challenges of doing research in a translocal field.

METHODOLOGY: A MOBILE ETHNOGRAPHY OF TRANSLOCAL TRADING PRACTICES

For a long time, 'the field' has been a rather unquestioned and uncritical dimension of geographical research. Nevertheless, over the last decades, especially due to an increasing interest in qualitative methodologies and the raising popularity of reflexivity and positionality (cf. Rabinow 1977, Rose 1997, England 1994, Cook et al. 2005, Nagar & Ali 2003), geographers have begun to critically examine the meaning of the field, fieldwork and the position and role of the researcher within the field. Discussions of these aspects in relation to my own proceedings will form the methodological context of the research. Developing contemporary trade as an empirical access to the translocal Swahili network and narrowing down my research focus, I will now show how my empirical experiences have driven me into a mobile ethnographic approach and what that means in practice. Positioning my research practices within current methodological debates in anthropology (where ethnographic research comes from) and geography (where ethnographic research finally seems to arrive), I argue that it is only mobile ethnographic research that allowed me to do the kind of empirical research enabling me to pursue the central aim of this book and thus write about the ways in which translocality is lived and a way of being in the world, and how this impacts on understandings of place and space.

SETTING THE FIELD THROUGH TRADE

> 'Our terrains aren't territories, they have weird borders, they are [...] rhizomes.'
> (Latour 1996: 46)

> 'In order to discern any "thing", we must distinguish that which we attend
> from what we ignore.' (Zerubavel 191: 1)

The field of research is not pregiven but has to be chosen and defined – and therefore actively created – by the researcher according to the research interests. Generally, practical reasons such as financial means, individual preferences and time constraints also play a role, though for a long time were tried to be hidden behind criteria, which seemingly derive coercively out of the research topic. Defining the field seems easier with respect to topics that are related to a particular site, possibly a neighbourhood, a city or a region with a focus on what happens within it. Nevertheless, choices concerning the empirical focus, research subjects (interview partners etc.) or the time frame still need to be made by the researcher and are seldom clearly induced by the topic. With a stronger focus on mobility and translocality in much of recent geographical research, the model of the field as a separate entity in physical space, like an island, that can be accessed and researched as a whole, has widely lost its appeal. Instead of seeing the field simply as a point on a map, it is therefore more and more often regarded as a dynamic set of relations consisting of different actors, things and ideas as they are connected in space and time with no clear-cut, natural boundary. Thus, the researcher has to decide upon which elements to include and what to exclude in the empirical study in order to gain the desired insights and information. Accordingly, my first methodological step had to consist of defining a field that would allow me to explore the translocal Swahili connections empirically by corresponding to my thematic and theoretical interest in aspects of mobility and their effects on the constitution of place and space.

Attempting to define clear boundaries surrounding the translocal Swahili connections in order to get a clear-cut field of research has to fail when acknowledging its dynamic and metamorphic character. Swahili people today are widely dispersed throughout the world, and while the highest density is surely still to be found along the East African coasts of Kenya and Tanzania including the islands of Zanzibar, there are also considerable numbers living in the UAE and Oman, as well as in many different cities in the USA, Canada, the UK and other European countries. Migration, remigration, demographic development as well as processes of 'swahilisation' (e.g. through marriage) constantly change, expand and restrict the connections in its different parts and render any clear fencing impossible. Translocal connections do not just end at a specific point, 'they just get flimsier, more difficult to discern' (P. Miller 1998: 363), so that, as Inkpen et al. (2007: 3) point out, 'distinguishing "external" from "internal" [would be] a matter of operational definition rather than a real distinction'.

Although such an operational definition of field boundaries might make sense and be helpful in other cases, bordering translocal Swahili connections would necessarily be based on an at least temporal (for the time of the research) 'freezing' of the connections and their mobile and fluid characteristics. However, it is exactly the dynamic and processual character of these connections that is crucial to any understanding of the ways in which translocality is actually lived and experienced. Thus, defining the field of research by drawing a clear boundary around translocal Swahili connections in order to distinguish where empirical research has to be undertaken would fail to incorporate the core interest of this study. The research interest does not necessitate in empirically addressing translocal Swahili connections as a whole or reducing research to particular places, but instead relies on a more detailed examination of particular connections. Apart from allowing for a deeper insight into the ways in which translocality is lived, this focus on particular connections also helps to avoid a huge field that would neither practically nor thematically be feasible for this study. Instead of setting the field by defining where to do the research, in cases like this, I therefore suggest to first concentrate on how translocal connections can be accessed empirically.

It is hardly possible to see, perceive or even examine translocality as such. Instead, one has to gain access to translocality through the ways in which it expresses itself. As Bourdieu and Waquant argue, 'we may think of a field as a space within which an effect of field is exercised [...]. The limits of the field are situated at the point where the effects of the field cease.' (Bourdieu & Waquant 1992: 100) Regarding translocality in its expression and effects guides the view towards the practices that make up translocal connections. These practices include a neighbourhood conversation in Zanzibar about the academic success of the family's daughter in Toronto, the turning on of a stereo in Dar es Salaam which has been sent by a cousin from London, the communal watching in Mombasa of the DVD from the last big wedding in Muscat, the hour-long phone calls between Glasgow and Pemba of two separated lovers, as well as the longing of an old man in Mpanda to rejoin his family and dismiss the current emptiness of his house, and many others which stand for the manifold connections and flows that, in their intermeshings and interminglings create and constantly recreate translocal space. As Rajchman (2000: 58) points out, rather than covering 'the whole' by moving from a beginning to an end, empirical research should therefore start somewhere in the middle by focusing on the practices through which the field comes into being. Also Bebbington and Kothari argue that 'engaging ever more deeply with the thickness of parts of the network can be immensely revealing in uncovering logics that reach far more broadly across the network' (Bebbington & Kothari 2006: 863-864). Defining the field guided by the question of 'how' instead of 'where' the network can be approached, therefore, requires to choose a particular expression or effect of the network, such as the practices mentioned above, on which the empirical research can be concentrated. With trade playing a central role in the translocal Swahili network, contributing to its constitution as well as being one of its outcomes, I chose this form of material exchange to serve as my empirical access.

Material exchange has for a long time served as access to social relations, as can be seen for example in the anthropology of the gift following Mauss (1923/1924) or studies on embedded economies following Polanyi (1944). In the Swahili context, it is trade that has a special relevance to the ways in which social relations are formed and held together providing an insight into the mobility of people, objects and ideas and the ways in which they intermingle on their routes and affect material and imaginative geographies of space and place. As the brief sketch of Swahili scholarship on this issue in the introduction to this book has shown, trade needs to be understood as both an economic practice and as an important part of Swahili culture and identity. In this respect, it addresses the complex relationship between economy and culture, thus avoiding reducing translocal connections to either one or the other.

Swahili trade consists of manifold connections, which are created, maintained, reworked and become significant through the practices of the actors who constitute them (Bebbington & Kothari 2006: 849). Thus, it is first of all the trading *practices* itself that are of vital importance to the empirical analysis. In order to account for the everyday dimension of translocal relations, it is particularly the rather small-scale professional and even non-professional trade among ordinary Swahili people in and in-between different places that enables an understanding of how translocal connections are lived, represented and experienced by avoiding an elitist bias. Secondly, in contrast to solely centring on humans in the examination of translocality, as still mostly the case, taking trade as the empirical access to the translocal Swahili network helps to address the *material dimension* of translocal connections, turning the attention to how not only humans but also material objects play decisive roles in the creation of translocal connections and bringing to the fore the 'binding quality' of non-human flows (Bunell 2007). Trading connections are not made of and by traders alone, instead they consist of complex interconnections of people, objects, and ideas that need to be examined in order to understand how translocal connections work and what they consist of. In this respect, special attention also has to be paid to the material constraints and problems encountered in the organisation of trade acknowledging not only mobile flows and connections but also immobilities and disconnections. Furthermore, the material manifestations of trade in particular places (in public and private spheres) will be of interest giving an impression of the way in which translocal connections visibly (re)constitute place and space. Thirdly, dealing with the *imaginative dimension* of translocal connections will help me to get to grips with the cultural and economic meaning of trade in relation to negotiations of Swahili identity. Here, the central question remains how translocal trading connections influence an understanding of a translocal Swahili space and provide a basis for a translocal sense of home and belonging. Overall, empirically engaging with these three dimensions of contemporary Swahili trade offers a way to explore the mobility of people, things and ideas as they come together in the constitutions of a translocal space – a space that could not be grasped by relying on a more classical idea of the field as a territorial, topographical entity.

THE GAINS OF ETHNOGRAPHY

On the basis of these reflections on the field of research, it becomes clear that only a qualitative approach is able to empirically engage with the translocal Swahili trading connections and to address the practices, materialities and imaginations constituting them. Coming to terms with how translocality is lived, experienced and influences senses of place and space requires a complex and deep impression of what is actually happening and done, expressed and represented in everyday life. As I will argue in the following, it is ethnographic research that is therefore best suited for this endeavour as it aims at getting as close as possible to the ways in which people themselves experience and make sense of their lives.

Being aware that ethnography is still not very dominant in geography, I want to use this section to first engage in more detail with the central characteristics and methodological thoughts of ethnographic research, starting with the origins of ethnography in anthropology. Building on that, I will then discuss the role of ethnographic research in geography, explaining why I consider ethnography as a valuable and enriching method not only in respect to this study but also for geography more generally.

Classical ethnography in anthropology

'Our disciplinary concerns may alter, our genres may blur, our theories may come and go, but ethnography remains "the anthropologist's muse".'(Comaroff & Comaroff 2003: 153)

When referring to a research method, ethnography generally means an extended period of fieldwork in which the researcher tries to integrate as much as possible into a relatively small group of people, taking part, observing, listening and documenting in order to gain an insight and make sense of the lived, routinised and mundane practices and experiences of their everyday lives. It was Malinowski who, in *Argonauts of the Western Pacific* (1922), defined the goal of ethnographic research as 'grasping the native's point of view, his relation to life, to realise his vision of the world' by 'plung[ing] into the life of the natives' for an extended period of time (Malinowski 1922: 25). His elaborations on how to conduct anthropological data marked the decisive change in the shaping of fieldwork as, until then, it was still common for researchers to live among white men and rely on selected informants to gain one's knowledge about the culture under study (cf. Clifford 1997: 54, Roldan 1995). Based on his own experiences, Malinowski argued that the kind of concrete data that could be gathered by remaining this distance does not lead to an understanding of 'real native mentality or behaviour' (Malinowski 1922: 5). Instead, he calls for a research context with 'no other white men nearby', promoting an intimate engagement over at least a year that allows accounting for the ways in which customs and believes are actually carried out and lived (Malinowski 1922: 17-18). In contrast to reconstructing culture from the expressions of informants, Malinowski pointed out the discrepancies between the spoken word and actions, developing ethnography as a method which foregrounds direct participant observation and is therefore able to actively and critically engage with, as he calls it, routine as it is prescribed, routine as it is done, and how the routine is commented by natives (Malinowski 1992: 18). In the years following this publication, Malinowski successfully managed to institutionalise his perspective so that ethnographic fieldwork has come to be, "the basic constituting experience both of anthropologists and of anthropological knowledge" (Stocking 2001: 289). In this respect, ethnography is much more than a method, it 'can be an aesthetic happening, *jouissance*, bodily labour and knowledge through all the senses' (Marcus & Okely 2007: 359). Some anthropologists even go as far as to see the ethnographic experience as a kind of initiation, saying that it is through fieldwork that one becomes a 'real anthropologist' (Gupta & Ferguson 1997: 1, Clifford 1997: 82). And despite changing and diverging theoretical and epistemological convictions – namely the ongoing debates on anthropology as a hermeneutic or interpretative *vs.* anthropology as an analytic discipline (cf. Lofland 1995, Schweizer 1993, Stellrecht 1993) – and increasing organisational and time constraints, until today, ethnographic fieldwork is still the key characteristic of anthropological research.

A major concern of the ethnographic approach is to go beyond discourses and verbal expressions. By observing and participating in everyday practices the researcher tries to get a more complex idea of the ways of life of the researched, adding what is actually done to what can be said. In this respect, a central feature of ethnographic research is to distance itself from a rigid and schematic research design. Instead it aims at adapting as much as possible to the specific research context and letting oneself be guided by it. Remaining open and flexible throughout the research process is crucial in order to account for the richness, complexity, plurality and messiness of the world without being trapped in preconceived categorisations and imaginations (Hannerz 1992: 6-8, Clifford 1983: 131-132). As Megoran argues, social scientific research methods such as surveys, semi-structured interviews, oral histories, and focus groups produce data by 'creating particular controlled environments that are structured by power relations and discursive formats generally alien to everyday forms of interaction' (Megoran 2006: 626). On the contrary, ethnography is based on the assumption that it is through participating and integrating as well as possible into the field, that the researcher is able to build up ordinary, and familiar relationships which open the way to more intimate and sensitive information that are decisive in order to gain an understanding of the ways in which people interpret and experience the world (Bernard 2006, Lofland 1995). Ethnographic research, as Geertz has put it in his seminal essay on *Thick Description* (1973), constitutes a search for meaning, not for law, and is therefore grounded in the in-depth study and interpretation of ideas, feelings, practices, and lived experiences, constantly striving for a holistic understanding of particular lifeworlds.

Although ethnographic fieldwork has long been accused by other disciplines of sticking to 'mere description' (Gupta & Ferguson 1997: 1), along with the cultural turn and the increasing reception of qualitative approaches, it has recently gained popularity beyond anthropology and can meanwhile be regarded as the most enduring contribution of anthropology to the humanities and social sciences.

Ethnography in geography

In an interdisciplinary context, the term ethnography is used in different and often overlapping ways. Whereas anthropologists rather tend to talk of 'fieldwork' emphasising the experience as a whole, other disciplines often seem to reduce the meaning of ethnography to 'participant observation'. Furthermore, the word is used for the monographic representation of ethnographic fieldwork (Megoran 2006: 625). Generally referring to ethnography as mainly participant observation, possibly complemented by little structured interviews, conversations etc., ethnography increasingly occurs in the repertoire of qualitative methods in geographic textbooks (cf. Blunt et al. 2003, Clifford et al. 2010, Cloke et al. 2004, DeLyser et al. 2010, Limb & Dwyer 2001, Reuber & Pfaffenbach 2005). Nevertheless, against the background of the discipline's history, ethnographic research means a big step from the mainly quantitative spatial science ('Raumwissenschaft') of the 60s and 70s towards a genuinely qualitative approach that acknowledges the value of less systematic, more diffuse, in-depth information and experiences (Cope 2010). Despite qualitative methods having gained broader acceptance and dominating many areas of contemporary geographic research, as Thrift has pointed out, they are often based on a 'narrow range of skills' (Thrift 2000c: 3) and mainly stick to conventional approaches in which the role of genuine ethnographic research is still highly contested. As it seems, many geographers are still 'torn between a constructivist approach and a longing to convey a "real" sense of the field' (Crang 2005: 225). Albeit positivism now usually having a negative connotation, often standing for 'naïve objectivity' (Williams 1976: 200), the common assumption that 'observation should be unbiased and representative, and that theories should be logical and consistent both with one another, and with observation' (Law 2004: 16) is still audible along with calls for 'value-neutral' and 'objective' social scientific approaches. From this position, ethnography appears as an 'overly idiosyncratic and subjective exercise, too reliant on the proclivities and orientation of the ethnographer' and thus providing unreliable and specious data (Herbert 2000: 558). On the other hand, there has been extensive work illustrating that interpretation is necessary and central to all science (ethnographies of science) and that no method is able to securely and necessarily give information about how things 'really' are (see e.g. Clifford 1986, Law 2004: 9). It is in this context, acknowledging the impossibility of certainty (Descartes 1986 [1641]), that representations and reconstructions of the world have been brought to the fore in geographic research, and with it a stronger interest in the ways in which these representations and constructions are created, experienced and negotiated. Recognising knowledge as always situated and partial, ethnographies are seen to provide a valuable contribution, as they are especially attentive to emic (self-ascribed), instead of etic (research-ascribed) views and meanings.

In German-speaking geography, ethnographic accounts are still rare and ethnography is only slowly establishing a stronger position in the canon of geographers' methods. In this respect, it can be considered as an important step that Reuber and Pfaffenbach (2005: 122ff) include and engage with 'participant observation' in their book on methods in human geography. Although concerns about its 'scientificness' are still visible and ethnography is rather presented as an interesting supplement than as a fully-fledged method, they state that it is underrepresented in geography and earns more attention (Reuber & Pfaffenbach 2005: 123). In Anglophone human geography, at first sight, ethnography seems to have developed a much stronger position in the field of qualitative methods. As Crang and Cook (2007) point out, over the last years, numerous books on the use of qualitative methods in geography have been published (e.g. Blunt et al. 2003, Clifford et al. 2010, Cloke et al. 2004, DeLyser et al. 2010, Limb & Dwyer 2001) so that, compared to the early nineties when they felt that they constantly had to 'justify what [they] were doing to a mass of sceptics' (Crang & Cook 2007: viii), qualitative methods now seem to have become a 'new orthodoxy' in human geography (ibid. ix). In this respect, ethnography is mostly referred to as 'participant observation+', also encouraging the creative use of approaches that strive for data going beyond words such as visual and sonic material (cf. Pink 2007, 2009, Rose 2007). Ethnography in a more narrow anthropological interpretive sense of the term is dealt with a lot less and not in much detail.

An early, very rich and sweeping geographic example of participant observation in geography, that has surely inspired a number of students and scholars to follow its methodological path, has been Crang's article 'on the workplace geographies of display in a restaurant in Southeast England' (Crang 1994). Moreover, over the last years, a few articles have been written explicitly calling for the use of ethnography in geography (e.g. Hart 2004, Herbert 2000, Laurier & Philo 2006, Lees 2003, McHugh 2000, Megoran 2006, Tuan 2001a). By presenting counter-arguments to its most frequently expressed criticisms, Herbert (2000) for example argues *For Ethnography* as being 'uniquely useful [to geographers] in uncovering the processes and meanings of social life as it allows 'to explore the lived experiences in its richness and complexity' (Herbert 2000: 550). Megoran more specifically addresses political geographers, demonstrating that ethnography could 'enrich and vivify the growing, but somewhat repetitious body of scholarship' (Megoran 2006: 623). Another field of geographical research, in which ethnographic research has increasingly come to be regarded as extremely enriching, is migration studies and with it the themes of mobility and translocality. As Fielding already claimed in 1992, 'only ethnographic research can reveal the subtle details of the experience of migration' (Fielding 1992: 205). In the same vein McHugh (2000: 83) has pointed out the potential of ethnographies to go beyond conventional definitions and measures and instead ground and reveal 'the play of migration and mobility in spatiotemporal re-orderings and transformations' and, by dealing with people in motion, explore their attachments and connections in multiple places. Furthermore, only through sensitive participant observation and close relationships to 'the researched' the often ambiguous and ambivalent migratory experiences can be accessed avoiding a one-sided presentation of migration as either excitement and freedom or loss and rootlessness (McHugh 2000: 84). Ethnography is therefore considered to be crucial in order to gain a more holistic understanding of mobility, being attentive to how material and imaginative geographies intermingle and become visible in the everyday practices inherent to translocal connections (Jackson et al. 2004: 3). This thought also applies more generally, as a stronger focus on topics such as embodiment, performances and emotions, which is expressed in recent human geographic and especially cultural geographic publications, calls for ethnographic approaches due to their special emphasis on the unspeakable (Davies & Dwyer 2007).

As this discussion of the role of ethnographic research, which is mainly led in Anglophone geography, shows, there are good reasons to promote the use of ethnography. This does of course not mean, that all geographic research should be done ethnographically. Instead, it is argued that ethnography should be fully included in the geographical repertoire of methods, as – depending on the research interest – an ethnographic approach can be indispensable in order to understand our relationships to space and place.

Regarding the central geographical terms space and place, ethnographies seem to provide the opportunity to get to grips with how people actually live their lives spatially as well as how they make sense of that experience. Ethnographies of place are able to show how places are lived, actively engaged with and attributed with meaning and emotions. In this sense, as Cook and Crang point out, ethnography seeks 'to understand parts of the world as they are experienced and understood in the everyday lives of the people who actually "live them out"' (Cook & Crang 2007: 1).

The need for ethnography

Gedanken ohne Inhalt sind leer, Anschauungen ohne Begriffe sind blind. *(Kant, KdrV, B75)*

Even though it seems as if some of the central methodological ideas of ethnography as developed in anthropology are slowly but surely making their way into geography, in some parts of geography the wish to attend to the experiences and understandings of everyday lives of those who actually live them meets with considerable resistance. Especially in the German context, in recent years, there has been a distinct move towards a human geography strongly influenced by social theory – an attempt to finally free themselves from the longstanding accusations of pursuing a geography void of theory and seek closer relations with other social sciences – making it even more difficult to gain support for research methodologies deeply anchored in the humanities. After all, the basic idea of the interpretative approach of ethnography clearly contradicts the theoretical convictions currently dominant in the social sciences. These can generally be characterised by the fact that they regard theories and concepts as helpful devices that serve to guide the empirical research as well as its analysis - an idea that is at the moment most prominently and very frequently expressed in the term 'theoretically informed research'. Here, theories and concepts thus somehow serve as templates that make any view beyond them extremely difficult if not impossible. As Ingold has put it, 'the way to know the world, they [social scientists] say, is not to open oneself up to it, but rather to „grasp" it within a grid of concepts and categories (Ingold 2006: 18-19). On the contrary, ethnographic research that is deeply embedded in the humanities considers all 'fore-meanings', i.e. theoretical convictions and concepts, as a possible distraction that has to be questioned and revised constantly (cf. Heidegger 1927: 153, Gadamer 1960: 251-252). From this perspective,

> 'a theory, by its clarity and weight, tends to drive rival and complementary interpretations and explanatory sketches out of mind, with the result that the object of study – a human experience, which is almost always ambiguous and complex – turns into something schematic and etiolated. Indeed, in social science, a theory can be so highly structured that it seems to exist in its own right, to be almost "solid", and thus able to cast (paradoxically) a shadow over the phenomena it is intended to illuminate' (Tuan 1991: 686).

Avoiding the separation of interpretations from what is actually happening is thus regarded to be a major task of an interpretive approach. As Geertz (1973: 25) has so powerfully put it, 'the generality [the interpretations] contrive to achieve grows out of the delicacy of its distinctions, not the sweep of its abstractions'.

In response to this critical stance towards the use of theory in much of the 'theoretically-informed' social science research, social scientists have regularly accused ethnographers of pursuing naïve empiricism. However, as a closer look at this argument reveals, the main difference between the two approaches is not that one is more theoretical than the other, even though this is how many social scientists prefer to see it, but that they follow a different understanding of (the use of) theory. The aim of ethnography is not to rest on detailed empirical descriptions but to use them as the grounding from which to – rather playfully – engage with theories, question one's instruments and, by doing so, further refine theoretical reflections. In order to be able to do this, it is necessary to remain an openness and flexibility towards the ethnographic experiences, always giving the empirical material the chance to surprise oneself. While 'theories hover supportively in the background [it is] the complex phenomena themselves [that have to] occupy the front stage' (Tuan 1991: 686). Only this allows researchers to let themselves become guided by the specific empirical context instead of imposing on it a rather schematic research frame, which impossibly matches the actual complexity and processuality. Striving for this kind of ambiguity and complexity seems to be particularly important when dealing with translocality, being one of the many concepts appearing in recent academic discussions that seem far removed from actual experiences. Therefore, instead of aiming for general theorising from above, what a thorough engagement with translocality needs is to develop arguments that derive from the particular experiences in the field. Only ethnographic research can provide the necessary empirical grounding to enrich and refine the theoretical debate on translocality.

With regard to this specific study on translocality in the Swahili context, without long-term participant observation it seems impossible to gain a deep insight into the ways in which Swahili people themselves experience and make sense of their translocal lives. Moreover, observing and participating in everyday trading practices and, through the development of trust and intimacy, gaining a critical and more holistic understanding of the constant negotiations, practices and discourses involved offers empirical material that goes beyond how translocality is talked about or how informants want it to sound when giving interviews. Here, the use of ethnography also helps to avoid prior classification, such as, for example, a distinction of people in important and less important informants (as generally required when selecting interview partners), and instead, as McHugh has pointed out, serves to 'capture the varying tempos and rhythms of movement and connection illuminating implications for both people and places' (McHugh 2000: 72). Finally, with its unique focus on ‚the everyday' ethnography opens up a way to grasp the casual, mundane, sometimes even random and incidental character, of translocal trading practices that is so crucial when we want to get to grips with what translocality entails for those who live it.

'The important thing about [ethnographic] findings is their complex specificness, their circumstantiality. It is with the kind of material produced by long-term, mainly (though not exclusively) qualitative, highly participative, and almost obsessively find-comb field study [...] that the mega-concepts with which contemporary social science is afflicted – legitimacy, modernization, integration, conflict, charisma, structure,...meaning – can be given the sort of sensible actuality that makes it possible to think not only realistically and concretely *about* them, but what is more important, creatively and imaginatively *with* them.' (Geertz 1973: 23)

FIRST ENCOUNTERS WITH A TRANSLOCAL FIELD

'To make connections one needs not knowledge, certainty, or even ontology,
but rather a trust that something may come out, though one is not completely sure what.'
(Rajchman 2000: 7)

Gaining access is usually introduced as the first of three stages of ethnographic research that are commonly discussed in textbooks: gaining access, living and working among 'the researched' and writing up (cf. Bernard 2006, Cook & Crang 2007: 37). Nevertheless, the research process is hardly as straightforward as this structuring in three stages suggests and are often overlapping. Writing about my own experiences and procedures I wish to convey a more vivid impression of the construction of empirical insights, how they have informed my theoretical thoughts and how this comes together in the textual outcome of this research. What I will present in the following, however, is not meant as a 'confessional tale [that] reveal[s] how the research came into being, expose the human qualities of the field-worker, chronicle the researcher's shifting points of view during the fieldwork and writing phases of the research, and remind readers that the fieldwork process is imperfect but not fatally flawed' (Magolda 2000: 210). Instead, I simply consider it as important to at least strive for transparency by addressing how I gained access and built up my field, showing how a 'rhizomatic' approach has become translated into concrete research practice.

Gaining access is sometimes presented as an incisive and dramatic rupture in a researcher's life, as a step that needs to be carefully prepared and that means a significant cut to one's life before. Whereas this may be true and also unavoidable in certain research projects, gaining access can also develop a lot more subtly and casually, almost incidental or random. My 'entry-point' to contemporary translocal Swahili trading connections has been in Zanzibar. On the one hand, in relation to my research focus, this could be justified historically by emphasising the idea that Zanzibar has long been the centre of Swahili trading activities. On the other hand, this choice can be seen (more faithfully) in respect to my personal connections to Zanzibar and the contacts I have been able to establish in Zanzibar during several stays prior to my research.

During an intensive two-month language course at the State University of Zanzibar in 2003, I had been accommodated in a host family, and it is this family to which I returned when starting the empirical research for this project in the beginning of 2006. This family who I have been staying with for almost a year in total lives in a three-bedroom flat in Michenzani, an area of Zanzibar Town which is dominated by long, narrow, five-storey buildings, built by the government (with support of the GDR) as part of the 'Revolution's Socialist Experiment' in the 1960s and 1970s (cf. Myers 1994). Suleiman, the father, and his wife Latifa have been living there since they got married in 1998. That was when Suleiman got back from his stay in the UK and opened his first business with his best friend Majid. Suleiman was born in Mwera, a village northeast of Zanzibar Town, where his mother is still living today. After he had finished Form 4 at school (comparable to the German 'Realschulabschluss'), his friend Majid and him had taken the opportunity to get into the UK, at that time made easier when showing proof of one's membership of the Civic United Front (CUF), at that time the major oppositional party in Zanzibar. Living in Milton Keynes, a satellite town northwest of London, they worked in a factory for three years, until the officials found out that their residence permit was not valid anymore and sent them back. Back in Zanzibar, Majid and him first opened a shop in Kiponda, an area in Zanzibar Stown Town, where they sold original international Music CDs. A couple of years later, when the business of CD burning had grown and the profit was decreasing, Majid and Suleiman opened up an internet-café instead. This internet-cafe right at the Michenzani roundabout still exists and is now owned by Suleiman alone. In addition to that, Suleiman owns a shop in which he sells clothes that he mainly buys in Bangkok.

Short after starting his business career in Zanzibar, Suleiman also started a family by marrying Latifa. Latifa was born in Wete, the second largest town of Pemba, as the second of eleven children by the same mother and father (The father has also been married to another wife with whom he got another ten children). After she had finished primary school, Latifa was sent to live with one of her aunts to continue her education in Zanzibar. That is where she later met Suleiman and agreed to get married to him.

Fig. 3: Location of Zanzibar and Pemba

Fig. 4: My first host family in Zanzibar   *Fig. 5: Michenzani*

During my first stay in this family, Latifa and Suleiman only had one child (by now there are four), a daughter called Nutaila, who was three at that time and had just started going to the neighbouring Qur'an school. Unless the third bedroom was vacant, I generally shared a room with her and the housemaid, and sometimes also with some other female guests. Usually, Munir, one of Suleiman's nephews, who worked in Suleiman's internet-café and served as the male 'bodyguard' at home during the times when Suleiman was absent, inhabited the third bedroom. During the day, Munir was hardly to be seen in the flat and, closing the internet-café as late as 11pm, he generally came home short before midnight and went straight to bed. Since early 2007, when his older sister got married, Munir has moved back to live with his mother in Malindi next to Zanzibar's harbour, leaving the third bedroom first to me and later to Nutaila and her sister Nuru.

Although, over the years, Munir has turned into a close contact and provided me with valuable insights concerning my research, it was Latifa's younger brother Mahir who facilitated my gaining access into the field the most. Especially in the beginning, he was almost omnipresent, regularly picking me up on a Vespa borrowed from a friend, taking me around and introducing me to friends and relatives. At that time, he was working in Suleiman's shop in Kiponda, so that was where I spent a lot of my afternoons listening to narrations about life in Zanzibar and engaging in conversations with other shopkeepers and customers. Apart from improving my Kiswahili and becoming familiar with different kinds of Kiswahili (formal and more informal expressions known as *Kiswahili cha mtaani*), I also learned a lot about the meaning of key terms in Kiswahili such as *ustaarabu* (maybe best translated as fine manners, respectable behaviour), *utamaduni* (culture, closely linked to urbanity), *mila* and *desturi* (denoting customs, habits, conventions) (cf. Bromber 2006, Kresse 2007, Middleton 2004), went to weddings and other religious celebrations, celebrated birthdays and simply spent a lot of time within the family, listening to the latest gossip, what is liked and disliked and observing and participating in everyday practices. Especially Latifa introduced me to the Zanzibari fashion and made sure that I was dressed appropriately and up-to-date when going out. Through visiting people and spending the whole day with them, a common practice called *kushinda* in Kiswahili that entails to integrate into *their* day, I got immersed into the everyday rhythms and routines of the family and beyond. Becoming integrated into Suleiman's family and joining them in their activities, I got to know more and more people, not only in Zanzibar but also in other places. I was for example invited to visit Latifa's family in Pemba, stayed with Suleiman''s niece in Dar es Salaam, his cousin in Mombasa, and with further relatives in Dubai. These contacts and the impressions gained from them play an important role in this research and, although they have been widely expanded, they form the context in which the research interest of this study has been developed.

What struck me when reflecting upon my first observations was the enormous role of translocal connections in their everyday life, forging my interest in the ways in which translocality is actually lived. Many practices such as regular telephone calls, email exchange and online chats, journeys, visits and the transportation and exchange of goods seemed as strategies to overcome or avoid distance and connect people and places. Although the importance of trade for processes of identification, which is especially referred to in the historical literature on Swahili people, has so far received only little attention in respect to the present, its crucial relevance to these contemporary translocal connections and the way they are established and maintained was strongly visible. People seemed to be constantly on the move in order to buy, sell, exchange, pick up or deliver things and almost every visit seemed related to some kind of material exchange. The focus on trading practices, which guided my empirical research, is strongly informed by these first impressions about the ways in which professional but also less professional traders as well as ordinary people create and maintain translocal connections and how they use different forms of material exchange to (re)constitute senses of common identity and belonging.

Building on how I first gained access to the field, as illustrated above, I decided to concentrate my research on a number of Swahili families. In London, for example, I was able to build on my connections to members of a family, which I had developed as part of the research for my MA-dissertation on Swahili people in London, and it was their relatives I stayed with in Mombasa. In Dubai, I spent most of the time with Suleiman's relatives, whereas on the Tanzanian mainland, I mostly visited Latifa's extended family. These in part very intensive connections to Swahili people in different places were indeed very helpful and facilitated my access to a wide range of people. Nevertheless, what is important to note here is, that the immersion into families, on the one hand, enabled me to gain deep insights into the lives of its members and broadened my empirical access by providing me with contacts in different places – I was actually handed over from relative to relative. But, on the other hand, 'belonging' to particular families also made it more exclusionary as I simply could not be everybody's 'child' or 'sister'. Hence, the families on which I concentrate in my study are either closely related or quite distant (see Keshodkar 2004 on the limits of moving among different families in Zanzibar).

All these families are of Omani Arab descent, being able to trace their presence on the East African coast back to the mobility of their ancestors from the southeast of the Arabian Peninsula to what became known as the Swahili coast. While one family belongs to the Kilindini, one of the Twelve Tribes of Swahili in Mombasa (*miji ithnashara* or *thenashara taifa*), usually regarded as the 'indigenous elite' and 'true Swahili', in colonial times, members of the other families would rather have been classified as Arabs (cf. Kindy 1972, Strobel 1975, Willis 1993). In the following, I therefore want to give a brief explanation why I still chose the term Swahili in respect to my empirical context.

Numerous Swahili scholars have engaged in the question of who could and should be called Swahili (e.g. de Vere Allen 1993, Eastman 1971, Horton & Middleton 2000, Parkin 1994, Prins 1961, Salim 1973, Topan 2000, 2006). The term Swahili already occurs in the travel writings of Ibn Battuta (Battuta 2005 [1929]) concerning his journeys between 1325-1354, where it is used to refer to the East African coast (*sahil*, pl. *sawahil* meaning coast in Arabic). According to Kindy (1972: 47), the term Waswahili has then been employed by some 'Swahili' to explain themselves to foreign Arabs, including Swahili living along the Tanzanian coast, on Pemba and Zanzibar, in Mombasa and as far north as the Benadir coast. The prefix wa- demarcates the plural form of the first noun class in the Swahili language mainly used for people.

In this respect, the term generally refers to the 'urban Muslim community that emerged along the East African coast around 800 CE' (Kresse 2007: 36, see also Horton & Middleton 2000, Nurse & Spear 1985), clearly distinguishing them from later immigrants from the Arabian Peninsula, particularly from Oman and the Hadramawt, who were classified as Arabs (cf. Middleton 2003). As an ethnonym, Swahili has more often been applied to others than to name oneself and has been closely linked to normative and political interests (cf. Glassman 2011, Kresse 2007). It is thus generally agreed upon that considering someone as Swahili has meant different things at different times, depending on the particular context and perspective (cf. Kresse 2007: 38ff, Topan 2006).

According to the observation, that many of those who fit the above mentioned critierias would probably first consider themselves as a member of a city or region, for example, *Waamu* (a Swahili from Lamu), *Wapemba* (a Swahili from Pemba) or *Wazanzibari* (a Swahili from Zanzibar), Prins (1961: 11) has come to the conclusion that people are never Swahili and nothing else. This can also be turned around, saying that, besides being something else, they are also Swahili. Though being aware of the complexity and internal differentiations of the term Swahili, Caplan (2004) and Topan (2000), for example, have pointed out the common aspects characterising 'Swahiliness' such as Swahili language, Muslim religion and migration. These aspects indeed also apply to the descendants of more recent migrants from Oman to the East African coast. Despite internal categorisations and boundaries, what can clearly be observed is a far more comprehensive Swahili identity, based on a common language and culture, which also includes those who might formerly have been categorised as Arabs. Still talking about them as Arabs would put them together with millions of other Arabs who have nothing to do with East Africa, and using more specific terms such as Omani-Zanzibari would exclude all those who went beyond these two places. For these reasons, Swahili seems to be the most appropriate term when referring to the different people I lived with, observed and spoke to during my research. Beyond family ties, it is their identity as Swahili that plays an important role in the constitution of the diverse translocal connections and is thus crucial in respect to the research interest of this study.

An occupation in the trading business did not serve as a criterion for choosing these families. However, various forms of trading practices and material exchange formed an important part of the everyday life in all cases, providing me with fascinating first insights and useful starting points to embark from when delving into my research topic. As illustrated in respect to my earlier stays in Zanzibar, research practices such as joining shopkeeper in their shops (*kukaa dukani*) and visiting and immersing into the day of family members and friends (*kushinda*), represented my first ways of 'deep hanging out' (Wogan 2004) in order to slowly gain an understanding of the ways in which translocality is lived in the Swahili context, always attempting, as Law has put it, 'to remain attentive to the unknown knocking at the door' (Law 1992: 380).

While striving for 'thick descriptions' of the ways in which people organise and make sense of translocal trading practices, mobility soon emerged as a dominant aspect, not only as a topic, but especially as a practice which interfered with the conventional idea of participant observation. In my research context people seemed constantly being on the move. Coming back to Zanzibar after a period of absence, I found out that a lot of people had moved to different places and were no longer where I had left them. Visiting people in London, I was always impressed about the high frequency of visitors from abroad, packed luggage almost constantly occupying the corridors. Biographies often revealed very short time residencies and showed how mobility highly influenced their lives. As it is exactly this mobility that is at the core of my research interest, it soon became clear that I had to develop ways to become more mobile myself. Instead of refining my empirical research to conversations about mobility or only grasping mobility in particular places, I therefore decided to follow a more mobile approach that puts mobility to the fore both as the contents and concerning the way of doing ethnographic research.

MOBILITY AS METHOD

Over the last years, in different fields of the humanities and social sciences, researchers have argued that a translocal perspective requires to do empirical research in more than one place (multi-site research) or even to accompany the movement of people, objects and ideas (mobile research). This argument is based on the idea that a translocal field cannot be confined to a single place but consists of sites, which are connected to one another 'in such ways that the relationships between them are as important for this formulation as the relationships within them' (Hannerz 2003: 206). Translocal understandings of the world and especially the new attention to mobility therefore question central assumptions of classical ethnography and have even lead to a fear that this could mean 'la fin des terrains' (Agier 1997: 69) and the end of anthropological fieldwork. These concerns have already been discussed intensively especially in anthropology (e.g. Clifford 1997: 52-91, Gupta & Ferguson 1997), so that I will only briefly sketch the main arguments in order to position myself within the discussion. Being convinced that mobile ethnographic research is necessary when aiming at understanding how translocality is lived and experienced, I will then critically engage with the practical ideas on how to actually do that. These have so far mainly been developed in the recently emerging field of mobility studies. Finally, I will discuss how these ideas on how to grasp mobility empirically have been taken up in geography, pointing out what I consider to be the most enriching elements of mobile methodologies for current geographic research and this study in particular.

Current debates in anthropology

> If ethnography was originally designed for small communities – how to do an ethnography of translocal connections? (cf. Tsing 2005: xi)

A widely expressed concern with regard to ethnographic research is 'about the lack of fit between the problems raised by a mobile, changing, globalising world, on the one hand, and the resources provided by a method originally developed for studying supposedly small-scale societies, on the other' (Ferguson & Gupta 1997: 3). Going back to Malinowski, it has generally been agreed that anthropological fieldwork ideally consists of long-term participant observation within a clearly bounded field. Though Malinowski himself was mobile in his research, accompanying the *Argonauts of the Western Pacific* on their trips (1922), it is these characteristics that have become particularly emphasised leading to what Marcus and Okely (2007: 357) have called an 'idealised, caricatured model' of anthropological research. As Clifford (1997: 67) points out, 'fieldwork has always been a mix of institutionalised practices, of dwelling and travelling. But in the disciplinary idealisation of the "field" spatial practices of moving to and from, in and out, passing through, have tended to be subsumed by those of dwelling (rapport, initiation, familiarity)' (see Amselle (2002) for a critical account of this debate). This image now seems to be disrupted by an increasing conviction that societies cannot be viewed as closed and situated anymore.

Building Appadurai's observation that, 'as groups migrate, regroup in new locations, reconstruct their histories, and reconfigure their ethnic projects, the ethno in ethnography takes on a slippery, nonlocalised quality, to which the descriptive practices of anthropology will have to respond' (Appadurai 1996: 48), Comaroff and Comaroff, in their article on *Ethnography on an Awkward Scale* (2003: 152) even go as far as to ask if ethnography has become an impossibility. Nevertheless, whereas on the one hand, many anthropologists have distanced themselves from the idea of territorially fixed communities and stable localised cultures and instead support the view of an interconnected and more dynamic world, Ferguson and

Gupta (1997: 4) have pointed out that simultaneously, and especially in reaction to doubts expressed on anthropological fieldwork by other disciplines, some anthropologists commit themselves, maybe even stronger than before, to long term fieldwork in a single setting (for a recent anthropological discussion on *How Short Fieldwork Can Be?* see Marcus & Okely 2007). The interesting question arising from this apparent contradiction is how these two aspects can be convincingly combined so that it becomes possible to account for translocal connections and mobility through places while at the same time relying on a method that brings to the fore the in-depth study of particular places and the research context. In order to ethnographically examine the translocality of culture and the ways in which cultures travel, merge, disperse or intermingle, it therefore seems to be necessary to 'denaturalise Malinowski's model, rediscover it and open it for alternatives' (Gupta & Ferguson 1997: 25) and, by doing so, to go beyond the 'localising strategies of traditional ethnography' (Appadurai 1996: 52).

According to Spittler, who accompanied the Kel Ewey Tuarag on their caravans to explore the relationship between the leader and his camels in the Saharan desert, participant observation should get closer to the ways in which travellers and explorers got to know distant areas, people and cultures in former centuries (Spittler 1998, see also the idea of 'fieldwork as travel encounters' by Clifford 1997: 67). Probably the most famous in advocating a more mobile and especially a multi-sited ethnographic approach is Marcus (1986, 1989, 1995, 2009), who summarises the attempts to account for a mobile world by outlining six ways of creating more nuanced understandings of the global through being attentive to the multiple connections between different locales. These six 'follow the ...'-strategies entail to trace people, their lives, the things they make and interact with in their daily lives, as well as the things they say and write, including the discursive strategies informing what they say and write (Marcus 1995: 106-10).

These suggestions on how to research translocality and mobility are explicitly designed and understood as a response to empirical changes in the world. However, in many areas of the world mobility and translocality are not actually new phenomena – in relation to this study take for example the early connections and exchange across the Indian Ocean (Pearson 2003) – but they have come to the centre of interest as a result of a different way of seeing the world that has become more relevant and been actively promoted in the works of Appadurai (1996), Clifford (1997) and others promoting or following the 'mobility turn' (cf. Adey 2010, Urry 2007, Cresswell 2006). On the other hand, it can surely not be denied that migratory movements have extremely increased over the last years and that new technologies facilitating movement and connections over distance have turned mobility into an omnipresent issue. Therefore, what sceptics of mobile and multi-site ethnographic approaches criticise is not the increasing attention and relevance of mobility and translocal understandings. What they doubt is the assumption that mobile ethnographic accounts can provide as thick and contextualised descriptions as classical long-term ethnographic research in a single location. Hence, it is argued that through mobile or multi-sited research the researcher is not able to gain a deep insight into the context of discourses and practices which is considered as essential in order to make sense of and interpret the complex lifeworlds of 'the researched'. Although multi-sited fieldwork has been practiced increasingly since the 1980s, when it became popular to study migration at both the point of departure as well as of arrival, to many anthropologists, multi-sited fieldwork remains an oxymoron, as they question the idea that multiple sites can be studied intensively without compromising the criteria of depth and thickness. Some also bemoan that with more sites involved research stays in each site necessarily shorten, so that the research often becomes confined to more easily accessible informants and has to rely more heavily on interviews, preventing the researcher from gaining an ethnographic grasp of the entire field. As Hannerz states in relation to his research on journalists, 'this tends to be the nature of multi-site ethnography' (Hannerz 2003: 207): from a 'homebody abroad', the fieldworker seems to turn into a 'cosmopolitan visitor' (Clifford 1997).

I would certainly agree to this criticism concerning much of multi-sited research, which often turns into a comparative study and seldom goes beyond seeing the different sites as separate entities, which are connected to each other, but not understood in a genuinely relational sense. Even though multi-site research practices have been designed around chains, paths and juxtapositions, they nevertheless often stick to research *in* multiple sites instead of moving *through* and *between* them. Thus, most multi-sited approaches generally still research in places and remain to look at mobility from different positions without grasping mobility itself. In contrast, what I want to stress is a *mobile* approach, that involves the actual moving with, accompanying and joining in mobility. In this respect, I argue that mobile ethnographies do not necessarily suffer context and in-depth interpretation.

When researching translocal lives and the ways in which mobility is organised, lived and experienced, it is only by being mobile oneself, that one is able to get access to and understand the context of the people one studies (cf. Latour 1999: 21). In a translocal setting, people are attached to more than one place (and the space in-between), so that it would not be possible to grasp their lives in only one location. Not staying in one place for long and regularly being on the move, people, as well as objects and ideas, contribute to turning apparently dispersed sites into coherent fields, creating a specific context which can only be understood through translocal research practices (cf. Hannerz 2003: 210). In order to gain an understanding of the context people relate to on the move, 'thick descriptions' therefore have to be based on mobile ethnographies containing in-depth and contextually rich material in places and on the paths in-between. Instead of thinking of mobile researchers as rushing through different contexts of which they only get a short glimpse, it is particularly through their mobility that they get an insight into what context might actually mean in translocal situations.

Mobile methods in mobility studies

In the field of mobility studies there has been a lot less concern about the possible lack of context as a result of mobile research, as the central aim here has been to develop methods that record mobility. To a big extent, mobility research is still occupied with the patterning, timing and causation of face-to-face copresence, observing and engaging with mobile bodies and travel in order to capture what brings people together, when and how often (Sheller & Urry 2006: 217, Bærenholdt et al. 2004). Apart from a rather quantitative take on mobility, the increasing interest in its qualitative dimensions has recently directed investigations towards the question of how the ways in which people relate to their surroundings are altered by mobility. In this regard, it has become the aim to 'bring mobility into the research process' (Ricketts Hein et al., 2008, 1266), and actively access everyday lifeworlds through mobility. In the following, I will mainly draw from Sheller and Urry's discussion of *The New Mobilities Paradigm* (2006) as well as from a recently published state-of-the-art article on *Mobile Methodologies* (Ricketts Hein et al. 2008) in order to briefly outline the developments in mobile research practices over the last years by particularly engaging with their benefits, limitations and challenges, always against the background of the demands of mobile ethnographic research presented earlier.

First, I want to address the strand of mobile methods that 'use movement as part of the research approach itself, so that generally the researcher is mobile and thus either follows the subject through space, or makes the subject mobile for the purpose of the research' (Ricketts Hein et al. 2008: 1269). Here, participation in patterns of movement is central while conducting ethnographic research or interviews (Sheller & Urry 2006: 217). Kusenbach (2003), for example, in her research on the perceptions of local problems in urban neighbourhoods in Los Angeles and how they are expressed in daily activities and interactions, has established 'go along-interviews' which allow the researcher to observe everyday activities while they happen and to access their experiences, reflections and interpretation at the same time. Whereas, in natural 'go alongs' the researcher accompanies journeys that the researched would also undertake independently of the researcher (Kusenbach 2003: 463), research-induced walks have been used to benefit from the visual prompts along the way that encourage verbal elaborations on the living experience in place (cf. Anderson 2004, Hitchings & Jones 2004). While Ricketts Hein et al. still state that, despite these 'walking interviews' being increasingly practiced, so far only limited materials have been published (Ricketts Hein et al. 2008: 1277), the number of articles on mobile interviewing is now constantly increasing (see e.g. Brown & Durrheim 2009, Carpiano 2009, Hall et al. 2008). In *A Single Day's Walking*, Wylie (2005) has shown how landscape, place and identity can be understood through the practice of walking. As Lee and Ingold (2006) emphasise in their work on *Fieldwork on Foot*, by foregrounding the movements, such an approach tries to account for a relational understanding of places as being created and constantly recreated by the routes people take to and through them (see also Ingold & Lee 2008).

Moreover, though still being in an explorational and experimental phase, the augmenting number of mobile ethnographic accounts already provide an impression of the benefits and challenges of this approach in relation to a variety of topics (see e.g. Warner Wood (2001) on mobile ethnographies of ethnic art dealers, Laurier (2004) on ethnographies of drivers on the motorway or Peltonen (2007) on a 'global ethnography' of middle managers in transnational organisations). Nevertheless, when looking at the summaries of mobile methods in the two state-of-the-art articles mentioned above, it is striking that the majority of the approaches under discussion are not mobile themselves but instead attempt to get to grips with mobility through particular, rather static, research practices (see also Watts & Urry 2008: 867). Provoking the researched to keep 'time-space diaries' in which they record their movements and activities is one way of gaining access to movements without directly observing or participating in them (Thornton et al. 1997). In contrast to examining mobility as it happens, this focuses the research on movement as reflected in the diaries documented in a static situation afterwards. Especially in order to trace personal mobilities and related emotions, biographic interviews centring around objects of memory such as photographs and souvenirs are used (cf. Tolia-Kelly 2004). Similarly, communication technologies serve as a valuable source to trace actual mobility as expressed in online communications. In addition, they are used to explore imaginative and virtual mobilities through websites, web-based discussion forums and email interaction (cf. Germann Molz 2006, 2007, 2008). Concentrating on discourses, also poetic and literary expressions are interpreted to delve into the atmosphere of place and translocality (cf. Jazeel 2003, Nyman 2009). Moreover, aiming to bring together mobility and immobility in place, researching transfer points such as hotels and airports creates an understanding of how 'immobile' places are strongly linked to, foster particular forms of and are necessary for mobility (cf. Adey 2004, Hulme & Truch 2006, Kesselring 2009). In these studies, often the number of passengers, hotel guests and, more generally, the passers through, play an important role in emphasising the high quantity of mobile people moving frequently through places all over the world. For some, GIS and GPS are welcome technologies to map and record mobility (cf. Mateos & Fisher 2007, Adriansen & Nielsen 2005) as they enable the surveillance of travel paths from afar.

With regard to this second strand of 'mobile methods', it becomes clear that a number of different strategies are being developed and increasingly deployed to research mobility without requiring a mobile researcher, but instead relying on the accurate documentation and reflections of mobile practices by the researched (in time-space diaries, online forums etc). Accessing mobility through discourses, these studies are either confined to a specific dimension of mobility or tend to translate discourses directly into practices. In comparison, researching mobility as it actually happens, seems to request less from the researched, since they should ideally behave as 'naturally' as possible, but demands more of the researcher, particularly in respect to the logistics. Accompanying people on their ways is time consuming, expensive, and needs a lot of organisational effort. Moreover, getting the opportunity to follow the moves of the researched generally implies close contacts and trustful relationships creating a hospitable environment for the constant and continuous company of the researcher. Furthermore, deeply engaging in mobility requires flexibility as well as the ability to let yourself be drawn into the movement and 'to attune to the horizon and rhythms of the subjects' (Burawoy 2000: 4). Overall, this shows very well, that mobile ethnographic approaches depend on the same processes of 'letting go' and immersing as far as possible into the research context as classical anthropological fieldwork. Only by fulfilling the central characteristics of ethnographic research, mobile researchers are able to gain a more holistic understanding of mobility than provided by other 'mobile methods'.

Mobile approaches in geography

Compared to anthropology, where the increased attention to mobility and translocal practices is discussed by some as something that is significantly altering the research process and somehow unsettle or even threaten central assumptions of anthropological fieldwork, in geography this focus seems a lot less disruptive as it can be seen in line with earlier interests in human behaviour in time and space. Often mentioned in this respect, is Hägerstrand's 'time-geography' (cf. Hägerstrand 1967, 1970, 1985), which entailed to map the paths of people through time and space producing graphs of movement including two spatial dimensions and one temporal. These time geographic maps served as the basis to analyse patterns of time use and the time-space 'choreography' of the individual's existence at daily, yearly or lifetime (biographical) scales of observation (cf. Pred 1977, Thrift & Pred 1981). Furthermore, the 'human activity approach' ('Aktionsraumforschung') developing in the 1970es concentrated on the movements of people asking for what reasons individuals would go how far to fulfil particular needs (cf. Horton & Reynolds 1969, for a German example see Popp 1979). This empirical examination of individual behaviour was generally seen as a valuable addition to the central place theory (Christaller 1933) and has especially informed spatial planning. Besides, in the field of transport geography, geographers have been occupied with the transient movements of goods and people since the early twentieth century (cf. Hurst 1974). Closely attached to quantitative methods and planning, since the 1970es, this has especially included the study of the patterns of transport networks, transport nodes and terminals, traffic engineering and the provision of scheduled services and, although less so, analyses of the movement of commodities (Hay 2000: 855–856).

Several elements of this earlier geographical research on mobility are still informing current research on the use of time and patterns of movement (e.g. Heydenreich 2000, Kesselring 2009, Kramer 2005) and some have even explicitly tried to adopt these older approaches to recent theoretical discussions and think them further (e.g. Couclelis 2009, Kwan 2004, Pile & Thrift 1995). Nevertheless, much of this research has been based on are more mechanistic and functionalist understanding of mobility and has concentrated either on the documentation and representation of movement in maps or, in applied geography, on its effects for spatial planning. In mostly quantitative studies, mobility has often been reduced to pure movement, with only little attention to discourses on mobility and mobility as an experience and way of being in the world. In the meantime, particularly with regard to the latter, qualitative methods have come to the fore and, as they are generally becoming more present in geography, are now being adopted to 'mobile research'.

As Latham has stated in his article on *Research, Performance and Doing Human Geography* (Latham 2003: 2000), when 'pushed in the appropriate direction, there is no reason why these methods cannot be made to dance a little'. Whereas, in contrast to anthropology, locational pluralism and multi-sited research are quite common in geography, so far, compared to other disciplines engaging in qualitative methodologies, geographic research can often be characterised by a certain methodological monotony, inflexibility and stiffness, in which the spoken and written word is still dominant (Crang 2005: 229). Only recently, geographers have tried to more experimentally take up the innovative elements of ethnographic approaches and applying them in the engagement with mobility, while also acknowledging that 'the world speaks in many voices through many different types of things' (Whatmore 2003: 89-90, see e.g. Cook et al. 2004, Cook & Harrison 2007, Latham 2003, Kindon 2003, Braun & Laurier 2005, Laurier 2004, Laurier & Philo 2007, Wylie 2005). This is also fostered by a growing interest in issues of materiality, embodiment and the importance of place in respect to mobility, needing mobile methods to put this agenda into practice (cf. Ricketts Hein et al. 2008: 1279).

Mobile ethnographies, in particular, offer a unique opportunity to get to grips with mobility and its effects on place and space, not only quantitatively, but, particularly if qualitatively, not only by considering mobility as pure movement or discourses, but by taking it serious as an experience. Following a mobile ethnographic approach, researchers become able to accommodate mobile and translocal research contexts and to attune to the mobile lives of the researched and 'participate in their continual shift through time, place and relations with others' (Watts & Urry 2008: 867). Seeing mobility as 'a way of being in the world' (Cresswell 2006: 3), through accompanying flows of people, goods and ideas, translocal flows and connections can be captured while gaining crucial insights into the ways in which translocality is created, dealt with, negotiated, and lived, seeing it as a specific way of becoming and belonging. Translocal connections, which, though under different terminologies, have been the subject of geographical inquiry for a considerable time, as Kracauer (1995: 252) emphasises, 'are not constructed according to a plan, like a firmly established system of thought, instead [they have] no other purpose than to be there and to testify through [their] very existence to the interconnectedness of things'. Researching these meshworks and creating a 'sensitivity to the intertwinement of the elements of the manifold' (ibid.: 252) can successfully be derived from and built on mobile practices of tracking and tracing, unfolding and unpleating complex and heterogeneous associations. Furthermore, mobile approaches stress the idea of becoming and processuality, and by always incorporating a sense of surprise and unpredictability, they lead to a greater openness to what may come out of the research. Overall, I argue that by bringing to the fore the ways in which movement is made meaningful and how resulting ideologies of mobility as well as well as mobile practices are implicated in the production of place and space, they thus tackle some of the core questions of geography.

DOING MOBILE ETHNOGRAPHIC RESEARCH

In the case of this research, the first empirical impressions I gained proved to me that mobility was central to this study, not only as a research interest but also as a research practice, which it seemed, I had to adopt in order to be able to get to grips with mobility as it is practiced and experienced. If I wanted to participate in and observe the everyday lives of those Swahili people I got into contact with, I had to be mobile myself. Otherwise, my insights would remain limited to discourses and reflections about mobility as expressed in rather static situations, for example while sitting in a shop, in a kitchen, in a living room or on someone's *baraza*, a bench or veranda in front of the house, often integrated into the wall of the building, which serves as a common meeting place (cf. Kresse 2007: 73). From more static ways of 'deep hanging out', I therefore moved even deeper into the research context and began to participate in, according to my experiences, an even more important practice: *kuranda*, which entails to join in the movements of the researched and to accompany them on their everyday paths in and through places. Being interested in the ways in which people deal with and constitute translocality, instead of doing ethnographic research in a particular place, this 'mobile deep hanging out' got me into taking up a mobile methodology which involved moving along with the people, ideas and objects and to participate in their movements while conducting ethnographic research on Swahili trading practices.

Accompanying trade journeys

Movement and mobility are central to any trade journey. Although Braudel (1986: 145) has been convinced that researchers can easily get to know about the practices and thoughts of traders by simply studying their notes, correspondence and accounts, I am convinced that something will always be lacking in these stable and residential accounts, especially when it comes to the lived experience of trading endeavours. Only through 'being on it' and directly observing mobile traders and their various performances of trade – how they relate to places, objects and people on the move – one becomes able to take the 'doings' of actors seriously and deal with mobility as it is practised and experienced. Hence, mobile ethnographies of trade journeys have come to form a central part of my empirical research.

The first opportunity to accompany some people I knew on a trade journey arose when Mahir, the younger brother of my first host mother Latifa, and his cousin Ibrahim invited me to join them on their trip through the Tanzanian interior, travelling to Sumbawanga, Mpanda and Tabora. These are towns on the Tanzanian mainland with a considerable Swahili population, in part dating back to the establishment of trade routes into the African interior, in part a result of more recent migrations from the coast into the mainland mainly motivated by the search for a better income. Having relatives in these towns, this route had developed into a beneficent business trip for the young men of the family as their relatives were not only keen to buy goods from the coast but also served as middlemen to find other customers. Eager to get an impression of how such a trip was organised and undertaken, I happily took this opportunity and accompanied them on their journey through the Tanzanian mainland. The next chance to accompany a trade journey emerged when some people I knew in Zanzibar had booked their flights to Dubai on the same date as me, were staying in the same place, and let me accompany them on their shopping tours through the city. Since the late 1990s, Dubai has turned into one of the most attractive and highly frequented cities for Swahili traders and, due to the enormous demand of goods from Dubai, even trips which were originally primarily intended to visit friends and relatives on the Arabian Peninsula have turned into trade journeys, since bringing back and selling attractive goods provides a welcome additional income.

Beyond this economic dimension, the strong cultural and ideological relevance of trading practices, which struck me on these first two journeys, turned my interest towards trips which were not directly announced as business trips but still played an important role in connecting people and places over distance. I therefore took the invitation from members of a family I had become close to in London to travel to Mombasa with sixteen of them in order to celebrate the wedding of one of their sons and join them in their holidays. In this context, it became particularly evident how material exchange is not only a reason for translocal mobility but also one of its effects. Focussing on material exchange in more ordinary settings shows how particular goods are transferred in order to create the favoured sense of place or perform a desired sense of self and belonging, thus revealing a lot about the spatial imaginations in this translocal context.

These first opportunities, which allowed me to do mobile ethnographic research on three very different trade journeys, occurred more or less accidentally. However, the choice of routes depended on my relations to particular people, their views and opinions about the importance of particular connections as well as the accessibility of routes and places. Moreover, these as well as subsequent opportunities to accompany journeys depended on my communication, as I needed to be in touch at the right time to hear about trips and to be available and reachable to be told about them. As Hannerz has pointed out, the necessary choices are often made gradually and cumulatively, as new insights develop, as opportunities come into sight, and to some extent by chance (Hannerz 2003: 207). Thus, in contrast to searching for justifications of the necessary selection, I argue that, consequently following a mobile ethnographic approach, all one can do is to attentively follow the connections one manages to establish and to trust in their relevance. By doing so, I was able to gain 'deep insights' into the diverse encounters between people, things and places on the move, and through examining the everyday experiences and performances of traders I got access to the everyday dimensions of translocal mobility. I accompanied spontaneous and planned journeys, experienced and unexperienced traders, men and women, travelled in groups of many or only with a few, went on short and longer trips, with some traders attempting to make big business, others and already being happy about a small benefit; I got an insight into trade of a variety of goods such as cars, clothes, electronics or 'Arabic' items, such as incense. Concentrating on particular connections allowed me to gain an understanding of how links are created, how they are maintained, imagined, communicated and practiced and why they break apart, highlighting the significant amounts of energy, resources and organisation that go into sustaining translocal lifeworlds. What I want to point out here is that, by doing a mobile ethnography of trade journeys, I was especially able to engage with mobility as it is dealt with on the move, gaining an understanding of the practices as well as the discourses constituting and constantly reconstituting translocal trading connections.

Building on these lived experiences of translocality, I also took up other research practices, which attempted to tackle translocality from different angles and served to complement and cross-read my impressions.

Following object biographies

Apart from following traders on their journeys through different places, I also examined translocal connections by entering them through the travelling objects. During my stay in Tanzania in the first half of 2006, a particular mobile phone caught my interest when I became involved in the first processes of its adaptation in Zanzibar. Casually following its life course, I was soon struck by its mobility and when coming across the same mobile phone again in 2007, I decided to consider its biography as a promising way of looking at contemporary trading practices. As part of my ethnographic research, I engaged in conversations about this particular phone with its different owners and users and tried to trace and reconstruct its path in as much detail as possible. This strategy of 'following the thing' was particularly effective in tracing links among individuals and places by focusing on the different aspects that make the object move. Especially in rather messy situations, focusing on a particular commodity helped me to shed light on the practices involved in trading it, learning how it moved from one person to the next without getting lost in the multiple and constant mobilities of the traders themselves. Whereas accompanying trade journeys gave me a detailed idea of how the mobility of the traders is organised and experienced as well as how trading practices are arranged on a more general level, concentrating on single goods enabled me to get to know exactly how particular deals were arranged and what they meant to the people involved. Moreover, it turned my attention to the fluidity and ephemeral character of trade and material exchange by clearly indicating the short period of time that many objects stayed with their respective owners. Instead of seeing objects as a 'travelling companion' (Thrift 1995), in this case, different people are seen as the object's temporary travelling companions.

In this respect, taking a closer look at the biographies of particular trade goods opens up an important additional window through which to examine the translocal spatial and cultural practices in which the objects are embedded. Whereas, on the one hand, this means understanding the actual movement of an object and, with it, the relationships between people as well as the connections between places, on the other hand, it also entails developing an understanding of its mobility in the sense of its adaptability and the different ways in which it becomes appropriated, i.e. the change of the object itself (cf. Benfoughal 2002). Moreover, as Engström and Blackler (2005: 310) claim, 'it would be a mistake to assume that objects are "just given". [...] Objects are [not] constructed arbitrarily on the spot; objects have histories and built-in affordances, they resist and "bite-back"'. Tracing object biographies therefore seeks to account for the 'sensuous, concrete and polyvalent nature of commodities as they inhabit our daily lives' (Leslie & Reimer 1999). The variety of relationships between people and objects –despite their mobility – affirms their meaning as 'social glue' (Hill 2007: 72) and reinforces their active participation in the making and holding together of translocal connections regardless of distance (Pels et al. 2002: 11). In this respect, building on the ideas of Appadurai and Kopytoff (both 1986), also geographers have started to trace 'the forms, uses and trajectories of "things-in-motion"' as a means of exploring negotiations of culture, questions of mobility, translocal commercial networks and the complex connections between places (cf. Dwyer & Jackson 2003: 270). Tracing English Royal letters' journeys, Ogborn (2002) for example, has concentrated on the production, carriage and use of texts as material objects in order to understand their place in early modern trading networks. Another example of current cultural geographic work in this field, standing in the tradition of Mintz's study on sugar (Mintz 1986), is the work by Cook et al. (2004, 2006) on following food as well as Cook and Harrison's publication (2007) in which they follow West Indian pepper sauce. Bringing to the fore debates on 'conducting "follow the thing" geographies', Cook and Harrison present how 'in, and through, that bottle of sauce, an amazing array of complex connectivities and mobilities' seems to be being mobilized on different scales (Cook & Harrison 2007: 58, see also Freidberg 2001 *On The Trail of the Global Green Bean*). Studies like these and others have exemplified, how exploring the connections through which mobile objects move allows for a contextualisation of the object within prevailing cultural practices and to understand its utilities, meanings and symbolisms from within the frame of reference of those who interact with it (Bridge & Smith 2003: 259).

Throughout my fieldwork, I managed to closely follow the paths of a mobile phone, a headscarf, several dresses, a rice cooker and a stereo. By doing so, I was able to gain detailed impressions of the varying encounters between different owners, users, traders and these material objects. Furthermore, this empirical approach offered an insight into the ways in which object biographies refer to ideas of Swahili identity and how these are constantly being negotiated and (re)created through the engagement with these particular commodities. Overall, these object biographies illustrate that it is also the goods themselves that tell us a lot about the trading relations and networks, notions of mobility and the role of mobile trade in translocal Swahili connections.

Tracing material footprints of translocality

Accompanying trade journeys and object biographies has brought me to particular parts in different places, which are extremely influenced by translocal trade. Areas like Kariakoo in Dar es Salaam or Kiponda in Zanzibar with most shops being owned by Swahili traders convey a strong sense of the translocal Swahili connections. Whereas ruins and palaces still portray the material implications of the early trading relations (Sheriff 1971: 10) and serve as memorials of a glorious past, recent translocal trading practices are also extremely visible for example when looking at shops, businesses, Swahili material culture and interior design. Mobility always includes a certain situatedness, and seeing places 'as key sites in the sets of flows' (Featherstone et al. 2007: 383), as 'temporary placements for ever moving material and immanent geographies' (Amin 2004: 34), I tried to complement my mobile ethnographies by also considering some of the diverse material footprints of translocal trading connections. Similar to the studies of transfer points mentioned earlier, I concentrated on particular business areas in Zanzibar and Dar es Salaam, at a hotel in Dubai, and on specialised (business) travel agents in Zanzibar and London. Furthermore, I also focused on wedding halls in these different places as well as on private homes where especially the decoration and interior design indicate the high relevance of translocal material exchange in order to express Swahili identity and create feelings of home. Working with mappings and photographs, I tried to sketch central developments and characteristics of these places while also spending a lot of my time there, hanging out and talking in order to get a sense of place and the processes of (re)creating home and belonging over distance.

It was when walking along the streets scribbling down buildings and mumbling the names of the products on sale as well as the number of floors into a voice recorder that I felt most strongly as a researcher. Whereas other research practices were easily incorporated into everyday life, and rather emphasised a genuine interest into Swahili language and culture than a specific and complex research interest, finding the time to map and visually document certain areas could only be explained by research rationales and clearly marked me as a researcher. On the contrary, for example in wedding halls, depending on who I went with, I was generally perceived and labelled as Hanan's or Salma's 'sister', 'Nabila's best friend' or as 'one of Bi Sheikha's children', which all stressed personal relations as reasons for being included in the party and sounded much more favourable to me. Generally adapting to dress styles and the current fashion, and in particular contexts also wearing *buibui* and *mtandio* (long black robe and headscarf) brought me a lot of positive feedback and, although this is critically debated in academia and opinions about its appropriateness or necessity vary (cf. Gupta & Ferguson 1997: 205), it soon felt very familiar and comfortable to me and strongly contributed to a feeling of acceptance and belonging which was mutually expressed and helped me to distance myself from other '*wazungu*' (white people, Europeans). It was also appreciated with regard to the Islamic religion, and, on the one hand, resulted in treating me respectfully and less obtrusive (which I was happy about) but, on the other hand, also demanded of me to behave according to their Muslim culture, constraining for example the opportunities to spend time with and accompany male traders when being the only woman (which I was sometimes not very happy about). Nevertheless, while to a certain extent following the rules which generally apply to Swahili women of my age – though differing in their interpretation and implementation among different families and places- and being treated and expected to behave accordingly in my host families, I also got an exceptional treatment, was given more freedom and understanding, so that certain things were appropriate for me and justified by my non-Muslim background, my professional assignment (rather than my interest) and my being a guest (who needed to do and see certain things and places). Hence, my trips, such as the one through the Tanzanian mainland, were very welcomed, not only because they were part of my research, but also because it was highly appreciated that I would go to see and get to know their relatives and friends over there.

When being in people's homes, especially in places such as Sumbawanga, Mpanda and Tabora on the Tanzanian mainland, it became clear that some of these homes have served as important anchoring points for translocal trading practices for decades. Particularly people from the generation of Mahir and Ibrahim's grandparents, who have witnessed the development of these places for about 60 years and still live there, thus add an interesting dimension to the participant observation by recalling their experiences and memories.

Fig. 6: Traveller, disembarking from the ferry to Zanzibar

Fig. 7: The 'first Arab of Sumbawanga' telling his stories

Listening to life stories and following biographies on the Internet

Listening to the conversations of these old people and seeing the important position they still have today, I started to directly ask them to tell me their life stories in order to get an impression of their perception of the historical continuity and ruptures of translocal Swahili trade. Taking into account that biographical interviews do not allow for an accurate reconstruction of the past, my focus here was to gain an insight into the ways in which histories of trade are reconstructed and made sense of by the interviewees at a particular moment afterwards. As interpretations of the process by those who have experienced it, these oral histories provide individual understandings of the emergence and continuation of translocal connections and flows. Moreover, the personal narratives are not only about individual histories but also illustrate individual mobility and tell something about how practices, objects and ideas have moved through these connections across space and time. As Bebbington and Kothari (2006: 855) point out, 'these living memories can complement official and dominant explanations of change, challenge conventional discourses and interpretations, and offer alternative versions' (see also Nagar 1997). Furthermore, they add an important generational dimension to this study, showing similarities and differences between old and young people, old and young traders regarding contemporary trading connections and their views on the impact of these mobile practices on senses of place, home and belonging today. Being interested in the ways in which translocality is lived and a contemporary way of being in the world, this biographical approach also becomes relevant in respect to younger people. Their life stories, although spanning a much shorter period of time, illustrate how mobility emerges as a dominant theme, how it is created, experienced, and incorporated into processes of identification.

When, one afternoon, I was sitting in a London flat with my host sister Hanan and some of her cousins, engaging more in conversations with people who were not present than who were, I started to include the Internet as a special kind of transfer point into my research. Over the last years, a number of publications have dealt with the impact of the Internet and other communication technologies on ethnographic research (cf. Budke et al. 2004, Crang et al. 1999, Hine 2005, Miller & Slater 2000). The Internet serves as a crucial means to exchange news, present recent trips and discuss opinions, so that I used Facebook and instant messenger services as well as regular email conversations to follow biographies, movements and journeys and see how connections are practiced, discussed and represented in these through these communication technologies. Being included into their communications as a 'friend', I got access to their self-portrayals and communication and, apart from following their representations, I also benefited a lot from joining this dominant way of staying in touch when staying at university.

Often, the distinction between field and home rests on their spatial separation. In this regard, as Gupta and Ferguson (1997: 35) point out, 'to go into the "field" is to travel to another place with its own distinctive culture, to live there is to enter another world, and to come back from "the field" is to leave that world and arrive in this one – the one in which the academy is located'. When researching translocal Swahili trading connections and becoming part of them, defining field and home according to such physical, spatial criteria has not always been very convincing. Not being physically present is no criteria for being part of the connections as the central idea is to hold people, things and ideas together over distance. Over the last years, my level of immersion and belonging to the field has never been static; my attitudes and identifications with the field have varied and the quality and intensity of connections have differed, often independent of physical presence or absence. Whereas I was sometimes very content with the high level of trust and inclusion, at other times it felt that I was too much involved when I had to carefully navigate through family troubles. Nevertheless, I do not think that physical presence does not matter anymore when researching translocal connections, as – like any other ethnographic practice – virtual connections over distance strongly depend on closeness and trust, that need to be established beforehand in order to be recognised as part of the translocal field. Hence, I do not support the argument that, because of the continuous contact researchers are able to have with at least some of their contacts by the use of the phone and email, virtual communication can generally compensate for shorter research stays, leading to what Wulff (2002) has called 'yo-yo-fieldwork'. Still, I regard virtual ethnography as a crucial way to explore the different dimensions and facets of the everyday life of the researched, which also illustrates very well that the ruptures between being in the field and being at home are a lot less strong than often assumed. Furthermore, as much as online and mobile phone communication belongs to the lives of most researchers, it is also an important part of the lives of many Swahili and especially central to the translocal connections and mobility.

2ND HALT: RESEARCHING TRANSLOCALITY

As this chapter has illustrated, the empirical insights of this study derive out of a mobile ethnographic research of translocal Swahili trading practices. Contemporary trade with its strong socio-cultural relevance manifests itself in translocal connections, and can thus serve as an ideal empirical access to translocality. Attempting to get as close as possible to the ways in which Swahili people themselves experience and make sense of their lives, I have argued for an ethnographic approach that, by avoiding prior classifications and remaining a theoretical openness, strives for a critical and more holistic understanding of the ways in which translocal trading practices are organised, negotiated and effect (re)creations of place and space.

As I have highlighted, ethnographic research has as its aim to participate in and immerse into the research context in order to produce 'thick descriptions' of the lives of the researched. When I started my research, I practiced this 'deep hanging out' by integrating into my various host families and spending considerable time in shops (*kukaa dukani*) and with Swahili people in their homes (*kushinda*). In response to my first impressions, which clearly showed that the context of Swahili traders is more mobile than that, I decided that I needed to become more mobile myself if I wanted to grasp their lifeworlds and engage with the ways in which translocality is actually lived. From more static ways of ethnographic research, I therefore moved even deeper into the research context and adopted a, in my experiences, even more important practice: *kuranda*, entailing to join in the movements of the researched, to roam and drift. Aiming at using an empirical approach, which corresponds better to the everyday lives of the researched and is able to incorporate the mundaneness of mobility and translocal trading practices, I thereby took up a mobile methodology that can be considered as a 'mobile deep hanging out' which was then mixed with more active and purposeful mobile participant observations. With the central concern to research translocality as a lived experience, joining in mobility and accompanying the journeys of people and objects in and between Zanzibar, Mombasa, the Tanzanian mainland, Dubai and London forms the basis of this study. Building on the impressions gained from these mobile participant observations, I used a variety of empirical approaches, some of which being more mobile than others, to complement, expand or cross-read my insights and experiences. Applying mobile methods, in this respect, does not mean to travel around all the time and constantly overcome big distances. Instead, doing a mobile ethnography also entails to follow movements in places, in order to attend to how mobility is incorporated into everyday life even without travelling far away. Combining different ways of accompanying people, ideas and goods on their everyday paths in and through places proved extremely valuable in order to gain an understanding of how the complex and heterogeneous translocal connections are organised and how different forms of mobility are created, lived and experienced, how they affect places, senses of home and belonging.

OUTLOOK: FOUR KINDS OF SWAHILI TRADING CONNECTIONS

The central aim, which I pursued by taking up the metaphor of the rhizome as well as by following a mobile ethnographic approach – to foreground the dynamic, processual and complex character of translocal trading connections – also concerns the way of writing up and representing the research. As Cook et al. illustrate, although research is often tricky, messy, and an open-ended process, this is rarely expressed through the ways in which it is written up and presented to the public (Cook et al. 2005). So, how to account for the unruly experiences and sometimes even chaotic, confusing and contradictory empirical material except for fitting them into an authoritative, nice and neat representation? (cf. Crang & Cook 2007: 179, Clifford 1983: 120)

I am convinced that writing is not necessarily a process of separating and disentangling things in order to create a linear, consistent and harmonious story. Especially with regard to ethnographic research, letting the empirical material speak to the reader as directly as possible allows for illustrative accounts that convey the complexity, ambiguity and multilayeredness of particular situations. According to my aim to account for the dynamic and processual character of translocality in the Swahili context, the presentation of my empirical material will thus be arranged along detailed ethnographic stories that evolve along four different routes that I followed in my research. Moving along concrete observations by sticking to concrete journeys and connections, I try to avoid hasty abstractions and terminologies, which might evoke rather static and fixed conceptualisations, missing the often ambivalent and dialectic nature of ethnographic insights. The point is not to present several ideal types of trade journeys that might claim to talk about translocality in general, but, on the contrary, to go along specific stories to illustrate and explore the complexity and ambiguity of translocal lives and experiences.

Similar to the translocal Swahili connections also this text and the process of writing can be understood as consisting of manifold lines of thought which are not one-directional but form a constant back and forth, building on intermingling and interrelating ideas and fragments that can be put together and ordered in different ways. In this respect, instead of being reduced to a report on fieldwork as a set of fixed research findings, the textual representation in the form of four different kinds of Swahili trading connections is meant as a facilitator into the dynamic and situative character of translocality as a lived experience in the Swahili context.

Chapter	1 Finding one's way in(to) translocal connections: a trade journey through the Tanzanian hinterland	2 Living (up to) the translocal imagination: on and in-between business trips to Dubai	3 Sticking (to) trading connections: object biographies through the hands of wannabe traders	4 Searching for home in a translocal space: economic dimensions of translocal cultural practices
Connection	Zanzibar – Tanzanian mainland	Zanzibar – Dubai	UK/ Pemba – Dar es Salaam	London – Mombasa
Dimension of trade	Semi-professional	Professional	Ideological	Ordinary life
Type of mobility	Journey	Trip	Bustle	Visits and Relocations
Experience of mobility	Newness	Repetitivity	Steadiness	Necessity
Material forms	'Arabic' goods, mobile phones	Clothes, Textiles, 'Arabic' goods	(used) electronics, mobile phones	'Arabic'/'Swahili' goods, furniture, clothes
Imaginative geographies	Swahili/Arab-African	Swahili-Arab	'Proper' Swahili	Heterogeneity within 'Swahiliness'
Relationship of culture and economy	Family and Trade	Consumption	Trade as cultural ideology	Economy of cultural events

Fig.8: Four kinds of Swahili trading connections

An emerging theme, which is discussed from different angles in all these chapters, is the role of economy and culture and the various ways in which they intermingle and intertwine on the different translocal connections, showing not only *how* trade is affected by 'culture' but also *how* Swahili culture is shaped through trade. Each of the four chapters thus deals with a particular way in which culture and economy inform each other and, by doing so essentially contribute to the emergence of a translocal space.

The first chapter will take the reader on a trade journey through the Tanzanian hinterland. Accompanying two young men on their way from Zanzibar – via Dar es Salaam and Mbeya – to Sumbawanga, Mpanda and Tabora and back, I provide a detailed insight into how such a journey is organised and experienced, still staying within the national boundaries of Tanzania. From the acquisition and transport of goods to their sale, I show how small-scale trading practices offer a way for the two rather inexperienced traders to become part of the longstanding translocal connections between the East African coast and the interior. In this respect, the chapter particularly focuses on the historical dimension of this route, pointing to continuities and changes, emphasising how much of the current practices need to be understood in the light of memories of the past. The route of the journey is chosen according to the location of relatives and thus constantly involves the negotiation of family ties between these places in the interior and their relatives at the coast. Pointing out the crucial role of family ties for this kind of trade indicates how the whole journey has to be understood as a family visit as much as it is an economic activity.

The second chapter deals with trade on a larger scale, following Zanzibari traders on their business trips to Dubai. Although, on this route as well, family relations to Dubai play a significant role in the organisation of trade, especially concerning the selection and transportation of goods, the trip has a much more professional touch. A particular focus of this chapter lies on the periods of mobility and immobility characterising the everyday lives of the traders, as the large amount of imported goods also requires rather fixed spaces (stores and shops) to engage in their trade. Highlighting the material effects of these kinds of trading practices in famous shopping areas in Zanzibar and Dar es Salaam, I then shift the focus from the traders towards the consumers. By concentrating on decoration and clothing, this chapter points out how these material effects are always closely related to the symbolic meaning of the goods, indicating a particular translocal relationship not only of the traders but also of the consumers. Showing how translocal connections are discursively negotiated among traders as well as consumers, makes clear how engaging in trade is generally considered as an ideal career for a Swahili.

This high significance of trade in respect to Swahili culture and identity is taken up in more detail in the third chapter, which however focuses on the everyday struggles of 'wannabe traders'. Dealing with the large number of mainly young Swahili men from Pemba moving to Dar es Salaam to enhance their chances of earning a living through trade, I try to get to grips with their mobility within a much smaller space. Taking as an example the everyday live of four young men, who try to trade in goods sent to them by their relatives living in the United Kingdom, I provide an impression of the hopes and disappointments driving them through the city on their ways to find a customer. Following particular goods that pass the hands of these traders, it becomes evident how this trade can rather be characterised by spontaneity, irregularity and chance than by clear rational calculations as an economic reading of these practices would suggest. A cultural approach, on the other hand, hints at the high ideological value of trade in the Swahili context, providing the traders with a justification even in times of economic failure. In this respect, this chapter illustrates how specific practices such as storing and displaying goods, borrowing and lending are used to evoke a sense of trade also among ordinary people.

That a focus on the cultural dimension of trade should not completely neglect the economic rationalities involved in translocal connections is emphasised in the fourth chapter. Expanding the perspective towards material exchange in general, I take weddings as an event that illustrates well how economic considerations influence cultural practices as well as how cultural practices also encourage various forms of economic and material exchange. Concentrating on two couples, both coming from the same translocal background, I give a vivid portrayal of the different stages of the two weddings, one held in London, the other one in Mombasa, from the finding of a marriage partner, to the planning and celebration of the wedding, and finally, to the choice of where to live after the wedding. By pointing out the complex and frequent mobilities characterising the everyday life of the families involved, this chapter also aims at conveying a sense of what living translocality actually means to the ordinary people who are part of the translocal Swahili space.

MOVEMENTS

FINDING ONE'S WAY IN(TO) TRANSLOCAL CONNECTIONS: A TRADE JOURNEY THROUGH THE TANZANIAN HINTERLAND

As soon as the decision to go on a trade journey through the interior of Tanzania was definite, meaning that there were only a few days left until the departure, everyone involved in the journey was busy to get prepared as well as possible. For Ibrahim, who had already travelled along the same route twice, getting ready meant, first of all, to get hold of his warm jacket, which he had lent to a friend but would now need again to cope with the cold climate on the mainland in June. Being born and raised in Pemba and having spent the last three years living with his two older sisters in Zanzibar Town, where he was driving a public minibus (*daladala*), he had – apart from the two journeys – never left the islands. His cousin Mahir, who would accompany him for the first time, also grew up in Pemba and was meanwhile working in the shop of his brother-in-law in Kiponda, a busy shopping area near the market in Stone Town. Excited about the upcoming journey, he directly went to his brother-in-law to inform him about his absence and to ask him for some money, which would allow him to buy more goods to take on the journey. At that time, neither Ibrahim nor Mahir possessed any noteworthy capital, so both of them depended on relatives and friends lending them money in order to be able to buy goods. Since they were asking for rather small sums, it did not take them long to be successful. What seemed harder was the decision which goods to buy, not knowing for sure what would sell well and at such a good price that, when having to return the money, at least a small profit would be left for them.

THE SELECTION AND ACQUISITION OF GOODS

While Ibrahim and Mahir call their relatives living on the mainland, asking them for their opinion and announcing their visit, it is Ibrahim's two older sisters, Ruwaida and Rumaitha, who intensify the busy atmosphere by rushing to produce some *udi* (frankincense) that should be taken on the journey. Of course, this self-made *udi*, compared to the high quality products imported especially from Oman, is not the most profitable trade good, but it is nevertheless a welcome opportunity for the women to earn some money. Moreover, *udi* as well as body fat made of it are essential items for most Swahili women to live and perform their culture. By using *udi* and expressing and displaying this taste, they indicate to others a great deal about their socio-cultural position and their Swahili identity. Especially for their relatives living in the Tanzanian interior, with some of them never having been to the coast, let alone to Oman, these commodities from the coast play an important role in forming their imaginations of life in Zanzibar. In this respect, Ruwaida and Rumaitha also make sure that Mahir and Ibrahim take at least a few *buibui* (long black female overcoat worn by Muslim women), fabrics for dresses as well as shiny hair bands and clips, all being items that are more difficult to obtain on the mainland but serve as crucial expressions of their cultural belonging.

Knowing that they would not be able to deny taking these things completely, Ibrahim and Mahir try hard to at least minimise this 'female stuff' saving as much space in the luggage for more valuable goods. This is yet further constrained by the need to also take some culinary Swahili specialties to the relatives on the mainland. Mahir therefore goes to the market to buy some spices, octopus and dried anchovies (*dagaa*). Although on the Tanzanian mainland most ingredients used in Swahili cuisine can be obtained, their relatives will definitely expect them to take some of the specific Swahili delicacies for them, which are only obtainable at the coast. Similar expectations have to be met when visiting relatives and friends who live further away in places such as Dubai, London or the USA. Apart from the just mentioned, diverse fruits (*chokechoke*, mangoes, *mabungo*, *ubuyu* etc.) and vegetables (*kisamvu, mchicha, majimbi*) are heavily desired presents and are generally transported with great care even despite their weight. One of Mahir and Ibrahim's aunts, who lives in Muscat, regularly asks for octopus, and a friend's sister has just been sent a whole package of cassava leaves (*kisamvu*) to cook her birthday meal in Dallas. Even if some of these or at least comparable items can also be bought over there, it still means a lot to eat something coming from Zanzibar. Much more important than just being something to eat, it is the symbolic meaning, often strongly related to a sense of home and belonging, that makes these goods such a crucial component of Mahir and Ibrahim's luggage (cf. Slater 1997: 146).

In general, when preparing for the trade journey, Ibrahim and Mahir are extremely aware of the connotations of the different goods, so that their choices about which goods to take do not only refer to their economic position but also to the socio-cultural context and demands they wish to – but somehow also have to – represent. The frankincense, the Muslim overcoats and fabrics, as well as the selection of foodstuff all represent an essential part of Swahili culture with a strong emphasis on the Arabic links and identity, showing how the 'cultural baggage' of the traders is also strongly reflected in their luggage (Ogborn 2002: 156). But apart from the symbolic meaning of the goods, it is also very practical and financial considerations that need to be taken into account when trying to understand the particular constitution of this trade journey. All the goods filling Mahir and Ibrahim's luggage so far are goods, which can be acquired easily, without big effort and, at least some of them, at rather low costs. Furthermore, not being very experienced in this kind of trade, the young traders rely to a big extent on the suggestions of Ibrahim's sisters and other relatives who have been involved in these trading connections for many years. Soon, the pile of goods reaches the limit of what can be carried, just leaving some space for goods that Ibrahim and Mahir plan to buy in Dar es Salaam.

Dar es Salaam will be the first place we move through on our journey. From there, the plan is to travel to Sumbawanga first, move on to Mpanda, then go to Tabora, and travel back to Dar es Salaam and Zanzibar from there. This route clearly evolves from the location of their relatives, as in each of these places Ibrahim and Mahir have family members looking forward to accommodating them. The duration of each stay will mainly depend on the success of selling the goods, but we also agree to keep the timeframe of the journey to less than a month, so that neither Ibrahim nor Mahir would be absent from their regular jobs for too long and – similarly important – that I would still have time to pursue other issues during my six-month stay. Furthermore, especially Ibrahim does not seem to be too keen to travel through the cold interior for too long. But, though late May and early June will be the start of the cold season, they are also the driest months of the year in the area around Sumbawanga and Mpanda, which is important concerning the conditions of the roads – the chances are good that big overland busses are running from Sumbawanga to Mpanda, a service often suspended during the rainy season, as too many buses would get stuck or even fell over due to the muddy and holed road.

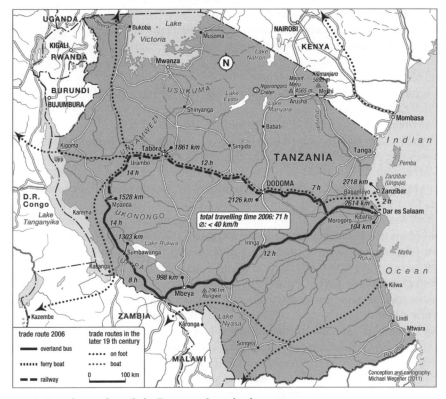

Fig. 9: Travel route through the Tanzanian hinterland

Having arrived in Dar es Salaam, it takes Mahir and Ibrahim a couple of days to buy some more trade goods, before taking off to the Tanzanian hinterland. With the help of some friends they mainly concentrate on the acquisition of mobile phones and some small electronics, such as a mini-stereo. With the demand of mobile phones growing rapidly on the Tanzanian mainland, they promise to be a profitable trade good on this journey. Depending on good offers, mobile phones can either be bought in Zanzibar or in Dar es Salaam and, having closer and more connections to people in Zanzibar, for Ibrahim and Mahir it would generally have been easier and more reliable to buy them there. However, it would also have been more risky, as random luggage checks at the harbour when arriving in Dar es Salaam might have led to complicated investigations about the acquisition of the phones, possibly ending in an accusation of theft or in having to pay high taxes or bribes. Especially traders with larger quantities of goods try hard to avoid paying this tax and often pretend to have bought the goods for personal use and not for commerce. Ibrahim and Mahir therefore rely on their connections to an old friend from Pemba who owns a mobile phone shop in Kariakoo, the main business area in Dar es Salaam. Going to his shop, they ask for phone models that are rather cheap, but not too simple, so that the phones would be a good choice to sell as 'first phones' to people in the interior with rather low spending power. As the shopkeeper has just received feedback from a woman who recently returned from a successful business trip to the Mbeya region, he confidently recommends an older model from Nokia, probably cheaply produced in Asia, of which Ibrahim and Mahir take four for approximately 60USD each. In addition, they also acquire two more fashionable, and also more expensive (about 130USD each) phones, which they consider as a good deal, though not being entirely sure if it would have made more sense to spend all their money on the cheaper ones.

After two days of preparations in Dar es Salaam, when having spent all the money, we are finally heading to the bus terminal in Ubungo to buy some tickets to Mbeya for the following day. Mbeya is the furthest we can go in one day on our way to Sumbawanga and Ibrahim keeps telling us how hard it will be to sit in the bus for a whole day. On the way to the bus terminal, we pass by at another friend, whom Mahir also knows from Pemba. He is repairing old computers, and having one of his computers connected to the Internet, he is also offering a kind of informal Internet café to his friends. More important to Ibrahim and Mahir is the fact that he is always up-to-date concerning the latest ringtones and famous songs that can be saved and played on mobile phones. Already in Zanzibar, Ibrahim and Mahir tried to store as many songs and ringtones as possible on their phones. While often acquiring new ringtones from friends for free, especially in areas where computer use and internet access is less common, you can take about 500-1000TSh for transferring them to another phone via Bluetooth, providing a welcome auxiliary income. Apart from the Bongo Flavour songs currently in the charts, some classics, RnB and Reggae, what is also important is to take some Arabic tunes and readings of the Qur'an as well as some Islamic sayings and images, which are particularly favoured by the Swahili community. Having added some more songs to their repertoire, and providing their friend with some songs in return, Ibrahim and Mahir now feel ready to depart.

The next morning we arrive at the Ubungo Bus terminal at 6am, as the bus is scheduled to leave at 6.30am. A cousin helps us with the luggage and we manage to get almost all our bags in the seating area of the bus, so that only the box with the foodstuff is stored in the boot below. Regarding the transport, another important aspect in respect to the choice of goods comes into play: their manageability. The goods not only need to fit the financial means of the traders and be appealing to possible consumers, they also need to be small and light enough to be carried and easily stored in a bus or train. What is important is that the traders themselves can take care of the luggage, as it often seems far too risky for them to give the things out of their hands, especially in crowded and busy places like bus terminals. Furthermore, the symbolic meanings of the goods are relevant not only in respect to the consumers but to the traders as well, as they also express their social and cultural context. In our case, Ibrahim and Mahir happily tell others about the mobile phones in our luggage, whereas the hair bands, female dresses (despite having about the same value as a phone) and *udi* put in my luggage are hardly mentioned. While in Zanzibar people had not been paying much attention to the anchovies and octopus in our luggage, already by the time we are entering the bus in Dar es Salaam, they have become a clear signifier of our place of origin and are met with positive remarks by the few fellow Swahili passengers, and rather doubtful looks by others.

Throughout the journey, the meaning of the goods will constantly become renegotiated in respect to the traders' views, the views ascribed to the potential buyers, and the views of their relatives in Zanzibar as well as in the interior, thus changing continuously according to our location. This indicates that, as Hudson has put it, 'the definition of a commodity, or of a thing, cannot be resolved by drifting off into the realm of floating signifiers, neither can its definition be simply and solidly anchored in given material properties, in contrast, the meanings of things, and things themselves, are stabilised or destabilised, negotiated or contested' (Hudson 2004: 457) always depending on their specific relations to people. Instead of being a solid substance congealed in its final form, material objects are in perpetual flux with their changing meanings being a crucial part of the making of social and cultural relations. This is particularly relevant, since an important dimension of this journey is the maintaining and strengthening of family ties, connecting family members living in the Tanzanian interior to their relatives at the coast, while at the same time signalling a much broader sense of Swahili identity and belonging. As Cook and Crang have already pointed out in their article on *The World on a Plate* (1996), knowledges about food's geographies are important in respect to understandings of self and other, here and there (Cook & Crang 1996: 137). Choosing goods which along the route will clearly be assigned to what is regarded as Swahili culture, this understanding will be traded along with these particular commodities, evoking pictures and memories of the coast, and emphasising a sense of Swahiliness among traders as well as consumers. Moreover, choosing rather prestigious goods in addition, such as mobile phones which are also of high value to the non-Swahili population on the Tanzanian mainland, further contributes to the self consciousness of the young traders in this to a big extent unknown terrain.

PASSING TRAVELLING TIME

> 'Without always knowing what we were doing, we were constantly adjusting to the arbitrariness by which we were surrounded.' (Naipaul 1979: 223)

Before dealing in more detail with the trading activities and how they are intertwined with family visits, I first want to attend to the traders' experience of travelling and passing through new landscapes and places. This experience plays an important role concerning the notion of mobile trading practices more generally. As a Swahili proverb says *kusafiri ndio kusoma* (travelling means studying) – it gives the traders the ability to see the landscape they know in relations to other places and it might even change the way they think by introducing them to different perspectives. Participating in mobile trading practices involves new experiences of distance and difference and therefore has a strong effect on the trader's lived experience of translocality. While on the one hand this trade journey connects Swahili in different places and brings them closer, on the other hand, it also increases the awareness of the great distances between the different locations.

Excitement

As soon as our bus has left Dar es Salaam to get on the TANZAM highway in the direction of Mbeya, Mahir starts to excitedly look out of the window absorbing as many of the new impressions as possible. For him, it is the first time to travel in such a big bus, and also the first time to travel on such big roads. To get a better view and to find out more about how it is like to drive such a large vehicle, he soon moves forward to sit next to the bus driver. Cautiously listening to his comments and stories about the route, he is fascinated by the life of long-distance travel. In his point of view, at the age of 25, it has been about time that he gets to travel and see new places, and finally getting this opportunity certainly adds to his self- esteem. On this journey, Mahir sees his first mountains and is very impressed by the changing landscape and vegetation. When passing through the Mikumi National Park Mahir and Ibrahim see their first elephants, giraffes, zebras and some antelopes. Although game parks and wildlife are probably the first associations most people in Western Europe have when thinking about Tanzania, these hardly play a role in most Tanzanians' lives and, also in the case of Mahir and Ibrahim, they represent a new image to be incorporated into their idea of the country. Getting an impression of the diversity and hugeness of Tanzania, it seems that Ibrahim and Mahir develop a certain pride, not only because they are able to travel and see all this, but also because they are part of this country – a feeling that is certainly odd to them, with a strong sense of Zanzibari identity and a clear averseness to the mainland normally being their dominant references. But this unusual appeal of the mainland – which especially refers to the environment – does not remain undisturbed as it is soon seen responsible for the cold climate. When taking the first shower in Mbeya and not seeing the need to wait until the water has been warmed up, Mahir uses the cold water eager to refresh himself after the long journey. A couple of minutes later we see him horrified – he would have never expected the water to be that cold – and despite wearing his warmest jumper and covering himself with a blanket, he does not stop shivering for the rest of the evening. Together with some of the impressions they get of the living conditions of people living along the route, these first experiences lead to lively renegotiations about the advantages and disadvantages of life in Zanzibar, overall turning the islands in a better light again. New experiences and their adventurous character remain a recurrent theme throughout the journey. From Mpanda to Tabora it is the first time for Mahir to enter the train, and although Ibrahim and I have independently taken this train on other journeys through the mainland before, the overnight journey in a crowded third class carriage without electricity is a new experience for all of us.

Fig. 10: Travelling by train

Ibrahim and I sit there drowsy but still trying to keep an eye on our luggage, especially during the numerous stays when people come running towards the train to sell their local products and some passengers use the opportunity to move their legs, get something to eat or at least some fresh air. Mahir, however, again takes the opportunity to get 'first class information' about the route and the train by making friends with the driver and sitting next to him in the locomotive. As Ibrahim finds Mahir's excitement a little embarrassing, thinking that he should appear more professional, he will later be a little mischievous when Mahir suffers from a cold that he got due to the cold wind blowing through the engine-drivers area. Nevertheless, both of them happily take pictures in front of the train in order to be able to give a vivid impression of their journey to the people in Zanzibar. (Unfortunately there was something wrong with the camera they had borrowed from a friend so that most of the pictures later turn out to be of bad quality.)

Involvement

Friends and relatives left behind in Zanzibar play an important role throughout the journey, not only by thinking about what to tell and show them after our return but, even more so, by involving them in the journey through mobile phone conversations.

In the bus as well as on train, about every five minutes Ibrahim and Mahir have a look at their mobile phone connectivity, exchanging SIM cards to compare when and where ZANTEL, BUZZ (now TIGO) and CELTEL (now ZAIN) still work. Not being reachable seems to be one of their worst fears, and they almost constantly stay in touch with friends at home, providing them with information about our trip. It's a shame I cannot hear the reactions of the person called when Ibrahim and Mahir excitedly state where they are and what they are doing in the Tanzanian interior, but I can definitely hear that it is a meaningful experience to the two of them. Each night, Ibrahim and Mahir make use of BUZZ' offer to have free calls after midnight, giving them more time to exchange news with their friends, tell them about the latest happenings or flirt with some girl.

Mobile phones have surely facilitated and intensified the connections between the relatives living in different places on the mainland and Ibrahim's and Mahir's family on the islands, both areas in which only very few people ever had a landline telephone connection. As visits are mostly induced by a family wedding, a funeral, or business arrangements – though the latter rather take place in Dar es Salaam and not necessarily lead to a visit of the islands – they only happen irregularly and do not always include the whole family, with the result that not everyone has yet met in person. In this context, the mobile phone has become a valuable tool not only to stay in touch but also to get in touch. Furthermore, particularly the use of SMS allows for a more casual (and less expensive) exchange of news.

For many scientists, mobile phones have gained much of their fascination from the idea that they decisively diminish the constraints of physical distance on social, economic and cultural life (see e.g. Graham 1998, Sanders 2008). Attention is paid, in particular, to virtual and imaginative travelling, ideally allowing us at any time to 'go' wherever we want. Most radically, this is announced by who Graham (1998) – in his discussion of recent writings on the relationship between information technologies, space and place – has called 'technological utopianists', who imply that the concepts of 'material space, place and time, and the body, [...] will be rendered obsolete', as 'communication technologies will be able to substitute for, and transcend, place-based, face-to-face interaction' (Graham 1998: 170-171). While Larsen et al. (2006) show, that communication technologies do indeed play an important role, enabling not only weak ties but also strong ties to be dispersed over space, they nevertheless also make clear that these are generally based on both recurrent long-distance communication and intermittent physical reunion. While some suspect that mobile phones replace a lot of physical travel and thus face-to-face contact, among others, they point to the possibility of virtual communication leading to physical reunion and, by reducing the need to travel, even supporting social cohesion and socialising activities at home (Larsen et al. 2006: 272, see also Sheller 2004). Moreover, surveys have shown that the majority of mobile phone calls and text messages have been sent over short distances of less than 50km (cf. Larsen et al. 2006: 273, Smoreda & Thomas 2001). This also holds true when looking at the call records, message inbox and outbox of Ibrahim, Mahir and some of their cousins and friends in Zanzibar. Calling their relatives on the mainland or abroad still remains something more special and serves a more general catching up – apart from when there is urgent news to exchange – and has not yet reached the more incidental, ephemeral and random character of many other communications to people nearby. Moreover, although it decisively contributes to being connected and keeping in touch with relatives and friends living in other places, the mobile phone surely does not replace travelling. The actual presence at weddings and funerals still remains crucial and also Mahir and Ibrahim's mobile trading practices could not take place without them actually going on this journey. Nevertheless, during these visits and while on the move, mobile phones at least allow them to somehow stay present in Zanzibar while being several hundred kilometres away (cf. Licoppe 2004).

Boredom

After the first excitement has passed and most friends have been called and told about the adventurous journey, Ibrahim and Mahir finally also get to experience the monotony involved in travelling over long distances. The seats become more and more uncomfortable and there is not enough space to stretch one's legs. Though mobility is often associated with speed and power – and the journey along the main road leading from Dar es Salaam to Malawi has indeed been rather fast, as every vehicle seems to tease out its limits –, when sitting in the bus from Sumbawanga to Mpanda or on a Tanzanian train, speed is certainly not an important criterion. Although the rainy season is over, some heavy rainfall caused the road between Sumbawanga and Mpanda to be extremely muddy and very reeled out. On this part of the journey, all of us constantly thrown back and forth in our seats. Sometimes the bus heavily hits the ground, sometimes the bus gets to lean to one side so much that we already anticipate the bus falling over completely. The bus is overcrowded and some passengers have to stand in the aisle for the whole twelve hours drive. With dense vegetation starting right next to the road, it is impossible to see very far and when it gets dark we just hold on to the seats in front of us and persevere until we finally reach Mpanda. Only the people standing next to our seats manage to keep their good mood and continuously joke around, making fun of the low speed and cheering for the bus to take us to our destination.

When at certain times of the year it becomes impossible for a big bus to take this road from Sumbawanga to Mpanda – and only a couple of jeeps secure minimal transport opportunities in that direction at very high costs – it is the train line to Tabora that becomes even more important to sustain the mobility of people and goods. This branch of the central railway line from Dar es Salaam to Kigoma reached Mpanda in 1949, forty-four years after it was started to be built in Dar es Salaam, and more than thirty-five years after the railway had reached Tabora and Kigoma. Three times a week it takes passengers the 333km from Mpanda to Tabora, a route that involves twelve stops and takes about fourteen hours. As one can probably imagine, although it is generally said that our contemporary mobility makes the world seem to shrink, this way of travelling, on the contrary, makes the world, or in this case Tanzania, seem a lot larger.

From Tabora it had usually been possible to travel to Dar es Salaam by train, but with railway works taking place near Dar es Salaam, when we were travelling in that direction all trains had to start and stop at Dodoma. Hence, when leaving Tabora to head back to the coast, we (only) have to sit in the train for nineteen stops from Tabora to reach Dodoma and then take the bus from there. Back on the tarmac road, our journey finally gains some speed again and Ibrahim and Mahir are clearly looking forward to reaching the coast. In comparison to the exhausting time spent on busses and trains, the boat taking us from Dar es Salaam back to Zanzibar, which often enough annoys its passengers by being delayed or driving at reduced speed due to some problems with the engine, feels comparatively relaxing.

Overall, this engagement with the act and idea of travelling shows how, on their journey through the Tanzanian mainland, the young traders experience mobility at different speed, different comforts, and different levels of excitement. Despite many experiences being new and adventurous, at certain times the journey also has a very mundane character. We simply become part of the flow of people travelling through the Tanzanian mainland, sometimes looking curiously out of the window, sometimes bored and tired, impatiently waiting to reach the destination. As Chaudhuri (1985: 11) has argued, 'the organisation of trade and the travelling, which reflect the scale of distances, reveal at the same time the mental world of the commercial community'. Although this journey only takes Ibrahim and Mahir through Tanzania, so they do not even have to cross any national boundaries, as I have shown, this route still offers a lot of contrasts and is full of strange elements. In the following, I thus want to focus on the effects of these new impressions and encounters on the 'mental world' of the traders. How do the experiences of the journey influence their 'imaginative geographies' (cf. Gregory 1995, Said 1993) of the Tanzanian mainland, their senses of belonging and attachment?

FACING 'AFRICA'? VIEWS FROM THE COAST

> 'We who lived [at the coast] were really people of the Indian Ocean. True Africa was a tour back.' (Naipaul 1979: 12)

The relationship between the islands of Zanzibar and the African mainland has for a long time been characterised by political tensions and cultural prejudices. According to Middleton (2003) in his historical ethnography of Swahili merchants in the early modern era, Swahili merchants saw themselves as the owners of *ustaarabu* and *utamadun'* (in that case translated as 'traditional wisdom based on long residence' and 'urbanity'), and what was outside their towns was called *ushenzi* ('barbarism', 'savage' or 'wilderness'; cf. Bromber 2006, Kresse 2007: 47). This dualism between *ustaarabu*, attributed to the Swahili in the inside of the town and *ushenzi*, used to refer to the wilderness outside of it, is very similar to the idea of *Orientalism* developed by Said (2003 [1978]). Here as well, Africa is presented as fundamentally different from, and essentially inferior to, the coast, as homogenous and unchanging. This strong dichotomy remains a prevalent opinion until today, frequently expressed in everyday conversations in Zanzibar and constantly fuelling the political tensions between the islands and the mainland. It is the idea of *ushenzi* that is also relevant when trying to understand the traders' engagement with the mainland of Tanzania.

Although the east African coast today is more fragmented and processes of migration have made the distinctions a lot less clear, the attitude of many Swahilis towards the Tanzanian mainland (*bara*) – as opposed to *visiwani (*on the islands) – and its population (*wabara*, as opposed to *waswahili*, or, in the Tanzanian case, more specifically, to *wazanzibari* and *wapemba*) is often still negative. In respect to Kenya, Kresse (2007) has illustrated well how Swahili today feel to 'have become downgraded to second class citizens; worse still, they have become ruled, and taken advantage of by the *wabara* […] who in the historical consciousness of Mombasa's urbanities used to be the underprivileged and less cultured outsiders, incomers, and, indeed, second-class citizens' (Kresse 2007: 76).

What is striking when travelling with Ibrahim and Mahir is their pronounced mistrust towards *wabara*. In general, *bara* is considered as being far more dangerous than the islands (as most negative aspects and especially criminal acts on the islands are often attributed to the *wabara* living there). While nobody of us was particularly concerned about our luggage at the harbour of Zanzibar, at all the bus and train stations on our journey, Ibrahim does not miss to remind us to be aware of thieves, often added by a remark on the bad mentality of the people around. Especially when arriving in Mbeya, the only place on the journey where Ibrahim and Mahir do not have any relatives, we try to get away from the bus stop as quickly as possible. Due to Mbeya's position right at the TANZAM highway and the TAZARA (Tanzanian Zambian Railway), the city is part of the most important connection between the harbour of Dar es Salaam and the landlocked countries Zambia and Malawi, and has therefore developed into a central transfer point. People from different directions with different destinations pass through and often have to stay a night, a fact that has led to the opening of a number of guesthouses right next to the bus station. Moreover, the bus station also attracts people who try to earn some money by carrying somebody's luggage or serving as an agent for a particular guesthouse or bus company. Especially at night, the place has become increasingly ill-reputed due to stories about people who have been mugged, had to observe or were even drawn into a fight. As hardly any of the travellers actually experience such incidents, it is especially the stories told about them that influence this negative sense of place. Seeing the bus stop as a threatening place, fits very well into the assumption elaborated by Malkki (1992) that a 'sendentarist metaphysics' dominates much of our thinking, resulting in seeing mobile people and places characterised by mobility as a threat to the rooted, moral and authentic existence of place (cf. Creswell 2006: 26-42). This case of Mbeya shows how mobility (of the others) becomes mistrusted and related to crime and violence, even if one is part of this mobility oneself.

When we arrive in Mbeya late in the evening, the first thing we do is to get a ticket to Sumbawanga for the following day. Although they are already sold out, we are able to still purchase three tickets whereas others get told to reschedule their journey. This is clearly because we are considered as 'Arabs' and Sumry is an 'Arab' bus company. When Swahili move to the Tanzanian interior, they generally turn/are turned into 'Arabs'. Though often Swahili-speaking, this particularly emphasises their Arab ancestry, and as not all Swahili living on the mainland have ever lived on the islands, it serves as the common denominator. Among themselves they generally know who directly moved to the mainland from Arabia and who is a *mpemba* or *mzanzibari*.

According to Ibrahim all Arab traders from the coast travel with Sumry. From previous journeys, Ibrahim knows the person in the ticket office who is a Zanzibari, and is proud to show us how well this 'Arab connection' works. On the whole journey we mainly trust in Arabs and especially in Mbeya we are suspicious of all others. Only Mahir, led by his curiosity, is a bit more open towards mainlanders, a behaviour that creates some hierarchical tensions between him and Ibrahim. After having bought the tickets, we then choose a guesthouse right next to the bus terminal, though not the one Ibrahim used to stay in on his previous journeys since that one is already full. Apart from going out to get some food, Ibrahim is keen to avoid any engagement with the people at the bus station, as he is extremely cautious not to get involved in any trouble. Besides, we are aware that we are considered as outsiders, supposed to be rich people from the coast who might be made to pay more. I, the *mzungu* (word used for Europeans), who is travelling with them, dressed and talking like a Zanzibari, surely increase the attention that we get. Nevertheless, I can observe a lot of similarities between the way it is for a European to arrive in Zanzibar and a Zanzibari to arrive in 'Africa', although they are still in the same country. (On another journey when we travelled from Arusha to Tabora through the Mbulu region, some Wairaq people even considered all three of us – Mahir, me and another trader from Pemba – as *wazungu*.) Whereas situations like this might even increase their feeling of superiority, this is often still mixed with the feeling of being a stranger, two sentiments constantly being negotiated on the journey, although the latter gets surpassed successfully in most situations by strongly relying on family and ethnic ties. These ties are mobilised at all stages of trade, and they are especially used in the process of finding customers and selling our goods as well as to help us finding our way around. Moreover, in some places, when needing to expand our connections to the local population, Ibrahim and Mahir extend their trust to some local Muslims.

Fig. 11: Meeting our local agent in Sumbawanga

Already on his previous trade journey Ibrahim got to know Mzee Adili, a Muslim barber in Sumbawanga, who helps us to sell two of our mobile phones and the mini stereo. This emphasises how, apart from family contacts, 'sporadic contacts that cross and link different groups' are vital to successful trading practices as they provide new and useful information and contacts (Grabher 2006: 176, see also Podolny & Baron 1997, Granovetter 1973, 1983). Selling relatively high priced goods such as mobile phones also means to always incur the risk of never receiving all the money, since customers usually want to pay in rates. In this respect, good connections are vital in order to successfully claim one's money. Nevertheless, on one occasion, Ibrahim had to stay in Sumbawanga for about two months to sell his goods despite all his connections. Instead of rational planning, what is often needed of the traders is a spontaneous, flexible, and situational handling of the diverse situations that occur on the journey. Although this can sometimes extremely increase the duration of the trade journey, this does not necessarily result in a closer engagement with the surrounding. The very selective engagement with 'the other' makes it possible to rather enforce than diminish the strong dichotomisation of 'the coast' and *bara*, *utamaduni* and *ushenzi* on this journey. Instead of 'facing Africa', Ibrahim and Mahir therefore rather pass through or even pass by 'Africa', enlivening the classical imaginative geographies and reviving the old image of the Swahili merchant.

THE PRESENCE OF HISTORY: SUMBAWANGA, MPANDA AND TABORA

When travelling through the Tanzanian mainland, one is not only continuously confronted with the current mobility of traders, but also with the long-term mobility of people and places along this route. Linking their material footprints to the imagined and remembered narratives of Swahili translocal practices and discourses, I therefore now want to illustrate how both crucially influence the contemporary paths and itineraries of connection as well as the sense of connectedness and belonging between these places.

Since 1820 and even more since the movement of the Sultan's residence from Oman to Zanzibar in the 1840es, Arabic traders and their caravans penetrated further and further into the interior of Africa. Soon, a distinction appeared between the traders who moved and settled in one of the halting places such as Tabora, Shinyanga and Ujiji and those who continued to move back and forth linking the trading routes in the interior of Africa to the maritime trade of the Indian Ocean (Bhacker 1992: 140). As Curtin (1984: 2) points out, the settling traders served as commercial and cultural brokers, 'helping and encouraging trade between the host society and people of their own origin who moved along the trade routes'. They generally stayed in constant touch with their relatives at the coast who provided them with supplies of furniture and other necessary items of the Arab culture (cf. Renault 1987: 12-14, Hahner-Herzog 1990: 13, Nicholls 1971: 74-100). As Middleton (2003: 521) has emphasised, the pattern of long distance exchange of the Swahili merchants was to a large extent based upon relations of trust, affinity, quasi-kinship and a common religiously validated morality.

With this image in mind, it is striking how little has changed and how much our journey resembles these journeys described in the literature: the visit of family members, relatives and Mzee Adili – who, even though not from the coast, at least still shares the same religion – as brokers to the local population, and the frankincense and dresses in our luggage to provide the relatives with the necessary 'Arabic' items. When thinking about the low financial position of Ibrahim and Mahir and their high aspirations, some comfort can even be drawn from Loarer's description of the *Population commercial de Zanguebar* (quoted in Sheriff 1971: 342), in which it says that, although most traders leaving with the caravans for the interior are 'the fairly poor Arabs who do not have a small piece of land and three or four hundred pounds of cloves', the new field was lucrative and many, after three or four journeys, acquired enough wealth to return to Zanzibar to establish or buy a clove plantation, which emerged as the most coveted niche for the Arabs (see also Nicholls 1971). Even if Ibrahim and Mahir surely do not strive for establishing a clove plantation anymore, this historical context is still vital when trying to understand the meaning of contemporary translocal trading practices. Translocality is not only about the here and there, but also attends to the then and now. And as the following impressions will show, this focus does open up our view on continuities as well as on ruptures, on stability as well as on loss, on unity as well as on tensions.

Meeting Mzee Mohamed in Sumbawanga

Through seemingly endless fields of sunflowers and maize we reach Sumbawanga in the afternoon of our second day on the road. The place is situated in Ufipa, a plateau between Lake Tanganyika and Lake Rukwa. Although not on the main and most famous caravan route from Ujiji to Bagamoyo, it still lies on the route from Katanga to the coast passing the southern end of Lake Tanganyika. It was most probably Wanyamwezi traders who first passed through Ufipa around 1820, perhaps even earlier (cf. Willis 1981: 82). Traders from the coast were definitely active in Ufipa from the 1850es onwards. As an heir apparent, Nandi Kapuufi in particular is said to have made alliances with coastal traders to support him in the fights against the Wabungu, and a little later in the early 1880es it is recorded that he has an Arab 'prime minister' (Iliffe 1979: 60, Willis 1981: 84). When in the early 20th century Arab traders increasingly suffer from the competition with Asian traders and move further into the peripheries, Sumbawanga is one of the places appearing to offer good business opportunities (Iliffe 1979: 141).

Today, Sumbawanga is the capital of the Rukwa region, with an estimated population of about 95 000 inhabitants in 2007. It is the host of the transport department for the region as well as several other smaller government agencies, and two conference centres are located in the city. More importantly, Sumbawanga serves as the major centre for commerce in the area, with the market – crucial for the food supply of the whole region – being situated right in the centre of the city. As there is hardly any industry in town, the economy is mainly based on agriculture and small businesses. Nevertheless, whereas originally the city was mainly inhabited by Wafipa, over the last years, also people from further away, such as, for example, Wasukuma, have moved there. However, with most of the Wafipa being peasants, and the Wasukuma especially investing in restaurants and bars, it is still Arabs who own most of the shops in town. Overall, Sumbawanga therefore has a small but stable Arab community and, since the place is continuously growing and promises a relatively lively business, it remains attractive to Arab traders if from the coast or other places on the Tanzanian mainland.

The bus stop is comparably calm, especially compared to our experiences in Mbeya, so we can unhurriedly look for a taxi. Just telling the driver the approximate direction is sufficient to be taken to the house of Ibrahim and Mahir's cousin Nasra, where she lives with her husband and their six daughters. According to the driver, 'everyone knows where the Arabs live'. Nasra is the oldest daughter of Ibrahim and Mahir's aunt Aza, a sister of their mothers. Aza, born in Pemba, grew up on the island and later, together with her husband, moved to the interior, where she stayed for more than twenty years. Only when her husband passed away, she gave up their house and shop in Mpanda to return to Pemba, where she is now living and looking after her mother.

Nasra was still born in Pemba, but spent her complete childhood in Mpanda. Then she got married to Rashid who was born and raised in Sumbawanga. Here, they now live in a spacious 5-bedroom house with a big courtyard, a little bit outside but still only a 10 to 15 minute-walk away from the city centre. According to Nasra, Sumbawanga is a friendly and lively place compared to Mpanda. Therefore, she is very content with her life, and although her own relatives all live far away, she has become included in the big family of her husband in Sumbawanga. As many Arabs on the mainland, Rashid has a transport business and is distributing goods to and from the surrounding villages. He often spends a couple of days in 'the bush' where, apart from doing business, he also engages in hunting. For dinner, we are therefore proudly served game, which is actually the first time for Mahir to eat this. After having passed on news and greetings from Pemba, Zanzibar and Dar es Salaam, Ibrahim and Mahir soon focus on the business. Rashid knows a lot of people, and telling about his recent experiences, he gives us an impression of what we can sell to whom. Nasra immediately takes over the responsibility for our 'female stuff', informing friends and relatives of our arrival and our goods, and selling the things for us. I soon figure out that it is considered impossible for Ibrahim and Mahir to advertise these female things, and even most of the food we brought is simply left with her, so that she has to arrange for it to be picked up. Already the next morning, the first women arrive who want to have a look at the fabrics, the frankincense and the hairbands. While Nasra tries to sell the things at a higher price than the one she has been told by Ibrahim and Mahir, so that she can earn a small provision herself, we go into town.

After a quick visit to the barber shop of Mzee Adili to show him the mobile phones and the mini stereo we want to sell, we directly go to the house of Rashid's father, who lives close to the market. From what I had heard, he is the most well-known Arab in town, so that we were expected to pay him our respect first. In front of his house cars are loaded with goods, supervised by one of Rashid's brothers. After the usual exchange of greetings, he tells us that they were already awaiting us inside but that we should not be shocked to see his father in a rather bad condition as he had just had a stroke a couple of weeks ago. Whereas Ibrahim and Mahir first spend some time with Rashid's brother, eager to hear about his business, I sit in the courtyard with his mother and some of her grandchildren. When Rashid's father wakes up, we all go in to see him.

Mzee Mohamed is considered to be the first Arab who settled in Sumbawanga. Although born in Kigoma, he was then raised in Oman and only came back to Tanzania in the 1940s looking for business opportunities. In the early 1950s he came to Sumbawanga by first taking the boat from Kigoma and then walking from Kasanga, a fishing village at the south shore of Lake Tanganyika. Since he speaks Kiswahili with a strong Arabic accent, it is not easy for me to understand him, but with the help of the others, Mzee Mohamed is excited to tell me about his experiences as an Arab trader in Sumbawanga. When he arrived in town, apart from the market there was not much more than some shanties and huts, and most of the building activities were only about to start. He therefore built his ample house close to the market with space for shops in the front of the house. Apart from opening a shop, he also started a transport business with several trucks and jeeps running between Sumbawanga and the surrounding villages. At that time, Mzee Mohamed spent most of his time sitting in front of his house, supervising the shop as well as the transport arrangements that had to be made. With more and more Arabs coming to live in Sumbawanga, all regularly passing by to greet him and exchange some news, the space in front of his house soon developed into a lively meeting place. To make these visits even more comfortable, he decided to open the first *mkahawa* (coffee house) in town. According to him, this place was what considerably strengthened the Arab community and made him famous. He proudly tells us that even the president of Tanzania comes to see him when he visits the region. Today, not much of this lively, future-oriented atmosphere that he evokes in his narrative seems to be left. The *mkahawa* is closed, and many of his old companions have left the town, most of them having returned to the Arabian Peninsula. Being about 90 years old, Mzee Mohamed has handed over his businesses to his sons, and now, since his stroke, he has not even been able to leave the house. Nevertheless, he seems to recover slowly, and with the help of his sons he does some daily walking exercise in the courtyard. Mzee Mohamed clearly intends to stay in Sumbawanga until the end of his life.

Fig. 12: At the old coffee house   *Fig. 13: Amongst grandchildren*

Even his wife agrees that she could not imagine leaving this place again, although she was born and grew up in Pemba. After her marriage, she also had to take the train to Kigoma and then a boat from there to Kasanga, from where she was picked up and taken to what seemed to her as 'the end of the world'. Although it took her a while to get adjusted to the new environment, she then made friends with the other Arab women and was soon busy enough with raising their children. Whereas their daughters all got married in different places, three of their sons still live in Sumbawanga. Like their father, they all married women from Pemba, with

Nasra being the only one who had already lived in the interior before. With altogether sixteen grandchildren in Sumbawanga there is still no time for loneliness.

Moreover, it seems to be these memories of past times that still motivate young traders today. As stated earlier, Sumbawanga has been growing steadily and, apart from an increasing presence of Wasukuma, as well as of people from Mbeya and Tukuyu, we also meet some 'Arabs' who have only recently arrived in the city hoping to find a way to make their living here. Zuhura, an Arab woman who owns a restaurant at the bus station for example, is just briefing her nephew on how to help her with her business when we pass by to say hi and pass on our regards from her relatives in Zanzibar. Most of the newcomers from the coast have either a relative or friend in Sumbawanga who has motivated them to come and usually provides them with a place to stay in the beginning and sometimes even with a job – often as a shopkeeper or driver. Many of them have not had a regular income where they were living before and put a lot of hope in this new opportunity. Whereas some do not stay for long, and some –like us – only pass through, others completely settle down, start a family and make themselves a home. While Mahir ponders on how his life would be like in this place, we walk back to spend the afternoon with Nasra and her daughters.

Already the next day, Mzee Adili tells us that he has found customers for two of the cheaper mobile phones and the mini stereo. Ibrahim and Mahir therefore follow him to finalise the deals. I instead accompany Nasra on her visits to a couple of friends and relatives to show them the 'female stuff'. Compared to Ibrahim's experiences on his last trip, we are rather successful in selling our goods quickly, and in order not to continue our trip completely empty handed, already on our third day we decide to take the next bus to Mpanda.

Madukani: The Arab Quarter in Mpanda

After a rather exhausting bus journey, we arrive in Mpanda around midnight. Again we take a taxi, asking the driver to bring us to the house of Mzee Sheikhani who is probably the most well known Arab in this town. The house in which we are going to stay is the one right next to his, formerly inhabited by Aza, Ibrahim and Mahir's aunt, Nasra's mother. When Aza returned to Pemba after the death of her husband, she left the house with Aunty Amira and her husband, who are not only living in and looking after the house, but also try to carry on their business. We still get served some tea before going to sleep and although it is already late at night, Aunty Amira is eager to receive news from Pemba and to hear about our 'adventures'.

The houses of Mzee Sheikhani and Aza are part of an area called *madukani* ('where the shops are' or 'at the shops'), which was built as a trading centre following the initiative of an Indian trader and financially supported by the Uruwira Goldfields Ltd. In the shape of a rectangle one house is built next to the other, all of them having a shop in front. Behind the shops there is the living space, with the rooms surrounding the courtyard in the 'typical' Swahili manner. Thus, already the style of the houses indicates who the inhabitants are and, indeed, today, this whole area is almost exclusively occupied by Arab traders. Most of them came to Mpanda after 1949 when the railway reached the town. Mpanda had only been established in 1937, when a mine was built and a settlement was needed for the miners. Thanks to the mining activities taking place immediately adjacent to the town, the place promised to be a profitable destination for the traders. According to the International Institute for Environment and Development (IIED), the Mpanda Mineral Field has lead-gold-silver mineralization in a number of places. From 1946 to 1960 copper ore worth of 5 million USD was extracted from the two main fields and, in addition, a considerable amount of gold was produced as a by-product. As a result, the population in Mpanda doubled between 1948 and 1957 (cf. Moeller 1971: 160). Nevertheless, the main Mkwamba mine was closed in 1960, although it is not even sure if the remaining deposit is not still large enough to allow for economically beneficial mining (Leader-Williams et al. 1996, Moeller 1971: 157). Despite small-scale artisanal gold mining still taking place in the area, the gold-rush-mood characterising the town in the 1950es, which has also attracted the Indian and Arab traders to move this far into the Miombo forest, has long past. (cf. Moeller 1971: 45). Today most of the Indian traders have left, and whereas in 1957 Mpanda is said to have had more than 10 000 inhabitants, ten years later in 1967 these were only supposed to be 2800. Today, Madukani seems to us like a ghost town.

We first visit Mzee Farouk, one of the old companions of Aza and her husband. Mzee Farouk came to Mpanda fifty years ago and tells us about the days when trade was more prosperous. He even speaks Kisukuma, the local language, and gives us an impression of how busy this area must have been, with the Indian and Arab shops surrounding the market which, at that time, was situated right in the middle of the rectangle. Peasants and miners from all over the region were coming to Madukani to buy whatever they needed. In this respect, Madukani clearly indicated the central position of the Arabs at that time. Whereas the Wasukuma had to rely on their instable stalls, selling local products and especially food, the Arabs were surrounding them with their large shops, overlooking the situation. The strong hierarchies are captured in the structure and infrastructure, in the clear segregation and the choice of particular built forms. Nevertheless, it is the same structure that today indicates the peripheral position of the Arabs.

Fig. 14: Behind the counter

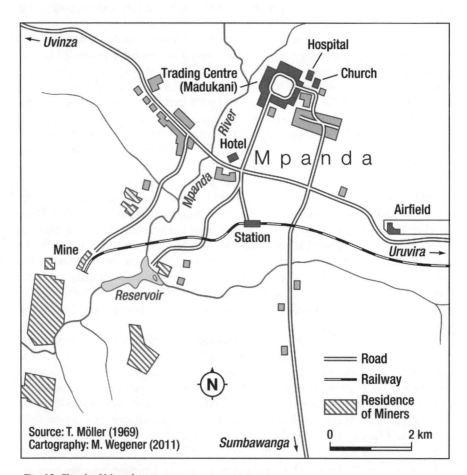

Fig. 15: Sketch of Mpanda

It was especially the movement of the market to the other side of town – which happened about ten years ago – that was like a stab-in-the-back. Once the vibrant town centre, it has now completely lost its busy character. Business takes place on the other side of town, where food and second-hand clothing is sold, and it is only the spacious structure of Madukani that hints at a more prosperous and powerful past.

Of around fifty Arab families that have been living in Mpanda, in many cases only the old generation is left. Most of the families have become related to each other through the intermarriages of their children, but, as Mzee Farouk tells us, especially during the last years, most of the young people have left to either Oman, the UAE or at least to Dar es Salaam or Tabora. Today, it is mainly the old ones such as Mzee Farouk and Mzee Sheikhani who carry on with their businesses, sometimes supported by their sons, who regularly travel to Dar es Salaam to buy new goods.

As our conversations show, the close translocal connections form an essential constituent of their lives in Mpanda. 'Cultivating their networks across space, and travelling back and forth in pursuit of their commercial ventures' (Portes et al. 1999: 225), however does not mean that traders do not progressively become integrated into 'local ways'. In contrast, the specific history of their translocal connections seems to tie them to particular places. So, even though the translocal Arab community in as well as beyond Mpanda might be the main reference for the old traders we spoke to, this does not diminish the attachment to the specific place in which they developed their businesses, but rather enforces it. Again and again, Mzee Farouk emphasises that he does not want to leave this place whatever happens. His attitude reminds me of Ibrahim and Mahir's grandfather in Pemba. Although the more prosperous times of his clove plantations have long passed and his bad health condition does not even allow him to leave the house anymore, he stubbornly fights against being taken to live with one of his daughters in Oman. Hardly receiving any medical treatment, the most important thing for him is to stay in the village in Pemba, where he had once established his career. Similarly, Mzee Farouk expresses how Mpanda is his 'adventure'; it is the place where he made it. And being one of those who went into the interior and independently set up their business, returning to the coast or to Arabia to depend on their relatives seems not to be an option. In this respect, attempts by Portes et al. (1999) to contrast assimilating immigrant entrepreneurs and translocal entrepreneurs, with the latter as the ones who refuse to be confined to either one space or the other, seem to idealise translocal traders as free-floating subjects. Mzee Farouk's urge to stay in Mpanda is even that strong that he still remains there when his wife is taken to hospital for a long-term treatment in Oman half a year after our visit.

This, as well as comparable situations among his neighbours, creates, on the one hand, a lonely and rather sad atmosphere in Madukani. On the other hand, this is also what seems to strengthen the cohesions and solidarity among the ones left behind. Moreover, everyone seems to appreciate our visit and interest in the place, showing them that they have not been forgotten.

When, on our walk through Madukani later that day, suddenly some camels appear and slowly walk through the empty area, the setting feels somehow surreal. As Aunty Amira tells us later, one of the Arabs imported some camels from Arabia 'just for fun'. Aunty Amira and her husband, our hosts, do not belong to this old generation of Arabs in Mpanda. Being about forty years old, they are one of the few 'in-betweens' and with their 6 months-old baby, they have just had the youngest Arab offspring in town. While Aunty Amira's husband frequently travels between Mpanda, Tabora and Dar es Salaam, she manages the shop, in which she is mainly selling *vitenge,* colourful cotton cloth mainly worn by the 'Africans' as well as some readymade dresses for children. She buys some of our frankincense and hairbands and displays them in her shop.

Fig. 16: Madukani *Fig. 17: The first encounter with camels*

For us, the remoteness and certain loneliness of the place turns into an advantage. As trade is decreasing, people travel at less frequent intervals to refill their shelves, and the supply is generally limited. One of the sons of Mzee Sheikhani is therefore happy to buy one of our more expensive phones, something he could otherwise only acquire when travelling at least to Tabora, and even there, most probably only at a higher price. Therefore, already after two days, we are content with our business and take the train to Tabora. After the quiet and calm surrounding in Mpanda, with not many young people around, Ibrahim and Mahir are especially looking forward to meeting some of their cousins living in this town in central Tanzania.

Tabora: Young careers following the old manner

> 'Towards the middle of this land, we came to a colony of Arab settlers and traders. Some of these had built excellent and spacious houses of sun-dried brick, and cultivated extensive gardens.' (Stanley 2005 [1909]: 257)

Tabora is the lynchpin of the railway lines through central Tanzania. The construction of the Central Railway Line reached Tabora in 1912, during the German Rule of Tanganyika that lasted from 1883 – 1919 (It was in 1893 when the German troops gained power over the town). The station building still serves as a reminder of the German past, welcoming us to the town when we arrive in the late morning. Having been an important administrative centre in colonial times, it remained to be the regional capital until today and is estimated to have about 150 000 inhabitants. Tabora is generally considered as one of the oldest urban settlements in the Tanzanian interior. Being situated at the junction of the three most important caravan routes leading through central Tanzania, Becker (1887) and Reichard (1892) assert that Arabs already settled in the region between 1810 and 1830. As a town, Tabora was founded in 1846 and soon developed into a lively centre of trade (Nicolini 2004, 2006).

Tabora soon formed a major entrepôt, with an increasing number of Arab and Indian residents, ruled by a representative of the Sultan of Zanzibar. In 1858, about twenty-five merchants are said to have lived in what Speke, one of the British explorers passing through the place at that time, called 'the great central slave and ivory merchants' depot' (Speke 1864: 97). While there was a core of permanent settlers – often those too old for distant travelling – there was a lager group who seasonally departed, or sent their agents, to distant frontiers. Throughout the mid-19th century, the town was the centre of the struggle for power between the Arabs and the Wanyamwezi, with their leader Mirambo finally winning over the Arabs in 1871. As Sheriff (1971) has pointed out, the community of coastal traders that grew up in Tabora mainly 'consisted of adventurers who had left the coast individually to seek or repair their fortunes in the interior, and they carried with them a strong sense of individual freedom, amounting almost to republicanism, which characterised Arabs of the time' (Sheriff 1971: 408).

The family we stay with in Tabora had not moved there directly from the coast, but came from Mpanda. The head of the family is Nassor, Aza's oldest son, who picks us up from the train station. He has left Mpanda thirteen years ago and moved to Tabora together with his wife Samira, a daughter of Mzee Sheikhani. Both had spent their childhood together as direct neighbours in Mpanda. It was when Saminra was sixteen that her parents considered her old enough to get married to Nassor. Now, they live not far from the old train station in Tabora in a big, manorial two-storey house. Whereas the ground floor contains an impressive living room, a bathroom and four spare rooms for guests, Samira, Nassor and their four children live upstairs. Surrounding the spacious courtyard, there are three more buildings, one containing the kitchen and a dining room, one containing three rooms, which were originally planned for the servants, and one building for the keeping of livestock. When we arrive, Samira and two female servants are cooking. They quickly put a mat (*busarti*) in the courtyard, so that we can sit next to them while still getting a fresh breeze. Nassor and Samira tell us about their experiences with the train. Despite the hassle, it is still the transport they usually use when visiting Mpanda or when travelling to the coast, but generally in the 1st and not in the 3rd class like we did.

Compared to the two places we have passed through before, Tabora already feels different, even though we have not yet been to the town centre. Whereas in Sumbawanga and Mpanda it had always been the historical dimension of Arab trading ventures that had been emphasised, and it often seemed as if the most prosperous times to establish one's life in those places have long passed, our first impressions of Tabora make the idea of Arab business endeavours on the mainland look a lot more topical. Overall, it definitely seems as if Nassor has managed to establish a rather stable and comfortable life here. But during our stay we also get to see how much effort this requires.

Nassor has recently become the head of the GAPCO depot. The Gulf Africa Petroleum Corporation (GAPCO) is one of the largest independent petroleum marketing and trading organisations with its main operations in East and Central Africa. Since 2007 the management control and majority shareholding has been taken over by Relience Industries Middle East (DMCC), registered in the United Arab Emirates. Besides working in the depot, he owns two trucks transporting petrol from Tanga to Tabora. As he says, these trucks have given him a hard time in the beginning. Almost on every trip one of them needed some repairs, got involved in an accident or was delayed due to the bad conditions of the road. Only with a big part of the road from Dodoma to Mwanza having been bituminised, road transport to and from Tabora has become a bit more relaxing, and more rewarding. However, to further diversify the sources of income and reduce the risk, his wife Samira tries to contribute to the family income as well. Only recently, she bought hundred chickens, which she now keeps in the third building surrounding the courtyard. While the male servant tries his best to rebuild the door, so that there remains a window but without giving the chicken a way to escape, it is Samira herself who is responsible for the feeding, the collection of eggs, and the cleaning of the henhouse. At least three times a day we find her in the middle of the chicken proudly calling us when she finds more eggs than the day before. According to her, this effort will hopefully soon be worth it, covering their own demand of eggs and allowing them to sell the rest to the neighbours and some street vendors.

Another business project Samira has just started is the opening of a shop. Together with her older sister Da Fatma, who has just moved to Tabora, they rent a space in *the* Arab shopping street in the town centre. Similar to Mpanda, in this road in the old part of town, all the shops seem to be owned by Arabs. Whereas the number of Indians has continuously decreased in Tabora and we only see a few big Indian shops near the market, the number of Arabs in towns seems rather stable. Some of the families owning the shops next to theirs have lived in Tabora for generations. Mainly selling materials and also some readymade dresses for children, Samira and Da Fatma take turns in shop keeping. Although it is still not clear if the shop is going to run well, they already sound very content with the project. They very much like the idea of having a shop in this place as 'this is what Arabs always do'. Sitting in the shop with them, it also becomes clear, that the shop in town offers a welcome alternative to staying in the house all day. Even if not everyone is actually buying something, it is a lively meeting place for chit-chatting, and when there is nobody to talk to, it is still interesting to watch the people passing by or to visit the neighbouring shops. The husband of Samira's closest friend Faidha – both grew up in Mpanda and then moved to Tabora with their husbands – also has a shop at the beginning of the road selling big packs of rice, flour, and sugar. The next day we are all invited for dinner at Faidha's place, an equally big house just one street away from Nassor and Samira.

What also soon becomes apparent is that the relative economic success of Nassor compared to many of his relatives brings with it an increased responsibility. Not only his mother – now living in Pemba – calls regularly to ask for some financial help for herself or her mother, but also some other members of the family rely on him. Hashid, for example, a cousin from Pemba, has come to live with him and his family and is now working in one of the GAPCO petrol stations in town. Also Said, the son of Da Fatma has been sent to live with them, long before Da Fatma herself has arrived, in order to enable him to finish his school in Tabora. Furthermore, a couple of months ago, Nassor's younger brother Najib has moved to Tabora with his wife Nayla, hoping to benefit from his connections. While we are there, he is working as the *msimamizi* (headman, overseer) of the same GAPCO petrol station as Hashid. Apart from Nayla, they all seem to have adjusted well to life in Tabora.

Having spent all her life in Zanzibar, Nayla does not find it easy to get used to life on the mainland. She already sticks out visually, as she still finds it hard to leave the house without wearing the long black overcoat (*buibui*), even though this is what the majority of the 'Arab' women living in the Tanzanian interior do. Nasra and the other women in Sumbawanga, Aunty Amira in Mpanda and now Samira, Da Fatma and Faidha in Tabora, they all wear long dresses or long skirts with a blouse and a scarf loosely thrown above their heads. In Zanzibar and Pemba this is only visible in the villages, and particularly in Zanzibar Town most women put considerable effort in looking 'Arab', represented by the fashion of one's *buibui* and the particular way of arranging the headscarf, generally imitating the latest styles from Dubai and Oman. This appears a lot less important in the interior. Due to lighter skin colour, different looks and speaking Kiswahili or even Arabic as their vernacular, the women are already externally identified as 'Arabs'. Here, all the women I meet state that they hardly wear a *buibui*, apart from attending special occasions or maybe when visiting people they do not know well. As a result, the selection of these long black overcoats in Tabora's shops is rather small and the designs are at least from the previous season if not older. However, they all want to make sure to have at least one fashionable one when travelling to Zanzibar or Pemba. That is why Faidha, who has been looking for a new *buibui* for a while, is happy to buy the last one we have in our luggage.

In general, within the 'Arab' communities on the mainland, dressing seems a lot less competitive than on the islands. It seems as if the women rather try to conjointly cope with the scarcity of the 'Arabic items' they favour and need. Samira and Da Fatma buy our last boxes of frankincense, some for themselves, some for sale in their shop. Concerning the fabrics we still have, they want to get a new skirt and a matching blouse made of it. Complaining about the bad tailors in Tabora who never know any new fashionable designs, they ask me if they can take one of my new dresses to the tailor as a sample. Generally, they wish to have a bigger choice of nice materials and designs, but with most of the local people wearing either readymade clothes or *vitenge*, the market seems too small for that. That is why a lot of the 'Arabic items' are sold privately. Everyone travelling to the coast seems to buy as much as possible and then invites people over or visits them to advertise and sell the goods. One day, a woman from the neighbourhood comes to show us *madira* – long, thin, colourful dresses made of cotton usually worn in the house or underneath the *buibui* – she has to offer; another day, I accompany Samira to see a woman who sells hairbands Samira wants to buy for her daughter.

Transfer points: enlivening historical imaginaries

Ibrahim, Mahir and I enjoy the days in this lively setting where we can observe directly how these young people arrange and build up their lives in the Tanzanian interior. Nevertheless, this positive orientation towards the future is also closely related to the past. On our last day before taking the train towards the coast, Nassor organises a family trip to a small museum in Kwihara, about four miles out of Tabora. A bumpy road takes us the last mile to an old Arab house. This house was the residence of Said bin Salim, the local governor appointed by the Sultan of Zanzibar in the mid to late 19th century. First having served as a governor's residence out of town, after 1871 – when Mirambo captured Tabora – Kwihara even became the main Arab settlement in the area for a short while. We walk around in this typical Arab merchants house, made of red clay with a quadrangle of large rooms built around a central courtyard. Signs indicate what the different rooms were used for: kitchen, bathroom, a stable for donkeys, and a room for the slaves. For most tourists the place is especially interesting as the place where David Livingstone had lived for a year in 1872 before he left for his final expedition to Lake Bangwelu. Also Stanley resided in this house in Kwihara before travelling to Lake Tanganyika to search for Livingstone. His companion Shaw died here in Kwihara and is buried in a field next to the house. A red sign on the wall in the front room indicates the place of his death. In the front rooms, there is an exhibition of several old documents related to Livingstone's stays and expeditions. When Nassor and Samira, Najib and Nayla, Faidha, Ibrahim, and Mahir look at the things and discuss their meaning, they clearly regard this history as the history of their ancestors. Despite the explicit references to the slave trade in the museum, they all show a bit of pride to belong to this group of Arab traders. Without being interested in the details, they create vivid images of the past times and link it to their life today.

Throughout this journey we have continuously been reminded of the long historical dimension of the translocal Swahili connections, with some of these connections dating back to the beginning of the 19th century. Material footprints such as the old Arab houses – especially visible in *madukani*, but also in the coffee place of Mzee Mohamed and the Arab shopping street in Tabora – well demonstrate the long history of 'Arab trade' in the interior. This becomes further illustrated by the narrations of a remembered and imagined past. Even more than the museum in Kwihara, it is the very rich life stories of the *wazee* (old, wise men), who have established their lives in these places and are now still there to see the young ones passing through, that link past and present. It is their and their ancestors' histories that serve as a driving force to enliven these and develop further connections today.

As passing through Sumbawanga, Mpanda and Tabora has shown, all these three places can be considered from different angles, the perspectives of long-term residents, the passers by and even from the one of the absentees. Nevertheless, whatever they are – the starting point for a new career, an Arab island in *ushenzi*, a business opportunity on-the-go, or a lonesome place to spend one's retirement –

in all three cases they are filled and refilled with the memories of a prosperous (and sometimes surely idealised) past. However, as the situation of Nassor and Samira, their friends and relatives in Tabora has shown, we meet much more than memories of the past: in a certain sense this history becomes re-enacted. Similar to the movements of Arabs into the interior in the 19th and the early to mid-20th century, even today we can still observe an increasing number of young men and women from the coast to move to the Tanzanian mainland to look for a profitable business. In this context, a trade journey like ours is crucial for the connections between the Swahili living in these different places. These are not merely places of origin or destination, but important constituents of a much wider translocal space. And although being situated in the periphery, they are still vital in respect to Swahili identity.

FACING THE COAST: VIEWS FROM THE MAINLAND

> 'Whoever has the daring necessary to face a journey into the interior, with its dangers, can, in a short space of time, become a rich man.' (Brode 2000:5, referring to early/mid-19th century)

As we have seen, for example, with regard to female fashion, along this route through the mainland, Zanzibar and the coast serve as constant references both for the mobile traders and for the Swahili residing on the mainland. While, I have earlier illustrated the particular relation between the coast and the mainland from the perspective of the coast, now looking into the opposite the direction helps to understand how Swahili in the Tanzanian interior position themselves and are positioned by their friends and relatives in Zanzibar. The kind of 'orientalism' underlying the perception of the mainland, although not directed at them, still influences the way they are seen by others as well as the way they see themselves. Thus, the view of the coast not only opens up an interesting perspective on senses of belonging and connectedness, but also on the tensions and hierarchies within this translocal space.

As Samira – who was born and has lived on the mainland all her life – states, her relationship to her family in law has long been a little tense. She is convinced that especially the women in the family would have preferred Nassor to marry a girl from Pemba. Although she feels more comfortable now than she did in the beginning of her marriage, she still has the impression that they are sometimes more critical and suspicious towards her than they would have been towards a girl from the coast like her sister-in-law Nayla. Generally, it seems that the discourse about the differences between Swahili *bara* and coastal Swahili are mainly led by women. In Zanzibar, women usually explain their critical views about their counterparts on the mainland by pointing to the fact that they lack a lot of things, which are central to Swahili culture and therefore cannot live like a 'proper Swahili'. Moreover, they accuse them of having been brought up differently (more loosely) as they do not live in a completely Muslim environment. Finally, it is often expressed that Swahili women who live in the Tanzanian hinterland are more (too?) self-confident. Overall, it soon shines through that it is a very schematic and generalised picture dominating the discussions. Nevertheless, it is also this idea that most Swahili women along the coast have in mind when they get married to someone from the mainland and consequently have to move there. Many are worried about moving to the mainland of which they have heard so many negative and strange things. Nayla as well, who has only recently arrived in Tabora with her husband Najib, tells me how strange the idea of moving to the mainland had been to her at first. She admits that she still misses Zanzibar but that she also starts to get used to her new setting, made a lot easier by her sister in law Samira and her family and friends. One year after our journey, Najib and Nayla, together with their newborn son, even move further away from the coast to Mwanza, a bigger city at shore of Lake Victoria, where Najib is offered a better position at GAPCO. While most of them need some time to get used to the idea of leaving the coast, engaging with their prejudices, when living on the mainland for a while, they usually soon change sides and then, equally generalising, critically look at and even distance themselves from the Swahili women in Zanzibar.

Women in Zanzibar are generally pictured as lazy and egoistic, obsessed with their looks, but not being able to do any hard work. Swahili women in the interior, on the contrary, consider themselves as being able to work hard and clench, bear things and not wasting so much time on dress, make-up and adornments (*mapambo*). Compared to Swahili women in Zanzibar, those living in the Tanzanian hinterland see themselves as being a lot tougher, having more strength and a stronger will. They ascribe themselves the power and courage to make their living in a strange environment and to familiarise with it. Although Zanzibar is still considered as the more beautiful place, it is argued that whoever wants his or her family to escape the bad living conditions faced on the islands needs to be able to sacrifice and take matters in one's own hands, i.e. to look for economic success on the mainland.

Concerning men, living on the mainland is generally reasoned by making an effort to improve one's economic situation. If they are not born there they have usually moved there for economic reasons as life seems more promising than at the coast and is often facilitated by the status of relatives who are already living there, as shown for example in the cases of Najib and Hashid. Especially young men from the coast either move to the mainland or organise trade journeys like Ibrahim and Mahir to benefit from the thriving markets in towns such as Tabora, Mwanza and Arusha. It is a way to actively try to improve one's condition, while referring to and living on the Swahili trader's way of life of the 19th century. From their perspective, those staying on the islands, constantly mourning about the bad political and economic situation seem rather lazy and spoiled, choosing to depend on their relatives in other places instead of making a move themselves.

Coming back to the idea of 'orientalism' and the ways of dealing with this strong dualism between coast and interior, it becomes evident that one has to differentiate different levels. Surely, the Swahili living in the Tanzanian interior do not totally assimilate into the local 'African' way of life, neither do they withdraw from their surrounding in any radical way, but they rather build on the long history of 'Arab traders'. As this discussion shows, there definitely is a distinction – and some tension – between Swahili in the Tanzanian hinterland and coastal Swahili, which is continuously fed by rather trivial and very generalised arguments. But this distinction is at the same time constantly renegotiated due to the mobility of people, which makes it difficult to clearly assign them to one category. Furthermore, on a higher level of abstraction, all of them still feel and cultivate a common sense of belonging and are strongly connected. And somehow they even mutually constitute each other, both representing the translocality so crucial to Swahili identity be it at the coast or elsewhere.

FAMILY AND TRADE: REFLECTIONS ON CULTURE AND ECONOMY (I)

Throughout the journey, it is striking how closely this trade journey is linked to translocal family ties so that it often seems as much a cultural event of a family reunion as a business trip. Ibrahim and Mahir have selected the route according to the locations of their relatives, and these connections can definitely be seen as a facilitator of the trading practices. Nasra and her family in Sumbawanga, Aunty Amira in Mpanda, as well as Nassor and his family in Tabora not only provide us with accommodation and food for free, they also decisively contribute to the success of business deals by advertising our goods to their friends, connecting us to possible customers or by simply buying the things themselves. Without being able to rely on these family connections Ibrahim and Mahir would hardly go on this trade journey at all. Nevertheless, while translocal family connections are vital to enable these trade journeys, translocal trade is also crucial for maintaining and strengthening family ties. As I have shown, many experiences and everyday activities are mainly family oriented: the visits of relatives, dinner invitations, small excursions and simply spending time together. Hence, despite trying to take and sell as many goods as possible, trade is only one dimension of journeys like ours. Realising this intertwinedness of family visits and trade makes clear that it is not solely an economic logic driving the organisation of these trade journeys, but that it is the close interplay of culture and economy that is crucial to understand the making and maintenance of translocal connections.

The issue of the relationship between culture and the economy, both ontologically and epistemologically, has been vividly discussed over the last two decades in a variety of disciplines, and also in geography (see e.g. Amin & Thrift 2000, Barnes 2001, 2005, Berndt & Boeckler 2008, Boeckler 2004, Castree 2004, Crang 1997, duGay & Pryke 2002, Gregson et al. 2001, Hudson 2004, Lash & Urry 1994, Martin & Sunley 2001, Ray & Sayer 1999, Sayer 1997, Thrift 2000b). After having regarded economy and culture for a long time like 'self' and 'other', each defined as what the other is not (cf. Crang 1997: 4), it seems as if they are now generally seen to be linked, somehow intertwined and even mutually constitutive. Triggered by the 'cultural turn', the core of the discussions, as Castree (2004) has shown, mainly lies in the critique of an, as Hudson (2004: 448) put it, 'often overly deterministic and structural reading of the economy and its geographies' that has long dominated the field of economic geography. Today, some therefore even propose to give up the distinction between economy and culture altogether and regard it as an 'entangled entity' (Amin & Thrift 2007: 145, see also duGay & Pryke 2002, Murdoch 2006). Others, however, take the opposite stance and point to the necessity of recognising them as separate entities (cf. Hudson 2004, Ray & Sayer 1999), while a third strand attempts to develop a compromise position 'maintaining that economy and culture are densely imbricated but by no means the same' (Castree 2004: 207). As Amin and Thrift (2007: 144) pointed out, 'how far these observations of culture-economy interpenetration have been allowed to shift theoretical fundamentals in different schools of economic theory is a moot point'. Instead of questioning and rethinking the basics of economic theory, it seems as if many simply 'added a cultural dimension to the conceptual package' (ibid.: 144).

One strand of 'cultural economy' now attempts to model culture as an economic input. Regarding consumption, for example, one of the classical economic practices, they try to look at how it is influenced by culture in the form of signs, symbols, desires and passion (see e.g. Baker 2000, Lury 2004). Another strand focuses on evaluating the economic potential of the 'cultural industries', recognising that culture – here mainly regarded as arts, media, leisure etc. – drives the economy. As it becomes clear in the recently published volume *The Cultural Economy* (Anheier & Isar 2008), in this respect, the field of 'cultural economy' is often understood as the economy of culture, or rather of cultural products. Sometimes even used synonymously with 'creative economy' or 'creative industries' it is mainly occupied with the economies of Art Festivals, material culture heritage, indigenous visual art, publishing, film-making etc. As the aim to '[examine] the relation between cultural economy and the rest of the economy' (Cunningham et al. 2008: 18) highlights, economy and culture are here still seen as something that can be completely separated, where cultural value can be seen against economic value. Overall, in both ways, culture is simply put into an economic logic, where on the one hand, the value of culture depends on its economic success (see e.g. Nyamjoh 2008 on the disrespect for cultural economy in Africa) and, on the other hand, culture itself is considered in economic terms, i.e. you can invest in culture, produce culture, sell and consume culture.

In contrast to these very stable and thing-like ideas of culture, a third strand in the field of 'cultural economy' turns the attention to an understanding of economic transactions as socially and culturally embedded (cf. Amin & Thrift 2007, Abolafia 1998). In this respect, each economic activity is seen to be embedded in 'webs of significance' (Geertz 1973: 4) through which people interpret their experience, and is thus always shaped by its social and cultural context. This builds on a long tradition particularly in economic anthropology and economic sociology, tracing the relevance of social and cultural aspects in economic interactions and practices of exchange (cf. Granovetter 1985, Plattner 1989, Polanyi 1944). In a more recent example, Stoller (2002, 2003) engages with the personal networks of West African street vendors in New York in respect to their economic interactions. In doing so, he shows that especially in the case of Nigerien Hausa, kinship and economic networks are inextricably linked, emphasising how the history of long-distance trade in West Africa influences contemporary trading activities in the United States (Stoller 2003: 206). And indeed, this is also the case when looking at contemporary Swahili trade journeys, where economic values are profoundly influenced by their cultural and family histories. Moreover, I would argue that the example of Ibrahim and Mahir's trade journey also shows that culture and economy cannot be clearly separated. Their journey is neither an economic endeavour nor a cultural venture, but it is the entanglement of both that gives it its meaning and makes it durable.

What this journey also illustrates is how trading activities have to be understood as a cultural practice (focusing on economic exchange) instead of seeing it solely through the eyes of classical or political economic theory. However, when reviewing the recent literature on this subject, it seems that so far mainly cultural geographers and scholars from the humanities more generally have been fervent in taking up this cultural reading of the economy, while economic geographers and economists rather argue for 'recovering a sense of political economy' and the engagement with 'the "big" socioeconomic questions' on a macro-level (Martin & Sunley 2001: 155-156, Jessop & Oosterlynk 2008). While a number of different understandings of the two key terms culture and economy are floating around in this debate, making it even more difficult to detect clear references and develop arguments, the central issue often seems to be to argue either in favour or against a thoroughly constructionist, qualitative and therefore somehow cultural reading of the economy, strongly influenced by the cultural and other more recent turns in the humanities and social sciences (see Amin & Thrift 2000, Barnes 2001, Rottenburg et al. 2000). In some parts of the discussion, it therefore does not seem to be the phenomenon itself but rather a more basic debate on epistemological approaches that is at stake and fuelling the discussions on 'cultural economy' in the discipline of geography.

As the emphasis on 'cultural economy' has mainly derived out of 'an unease with the economy as a layered, compartmentalised entity working to machinic rule' (Amin & Thrift 2007: 145), in what could be summed up as a forth strand, a strong emphasis is therefore put on teasing out 'the ways in which the "economy" is discursively as well as materially constructed, practiced and performed, exploring the ways in which economic life is built up, made up and assembled, from a range of disparate but always intensely cultural elements' (Hudson 2004: 454, see also Callon 1998, Law 2002, Murdoch 2006). As this journey shows this can only be achieved by following an ethnographic approach, since such an understanding of cultural economy 'has to be built up from an engagement with the lived social practices rather than deduced from macro characterisations of pre-given social moments' (Slater 2002: 61). In this respect, it is also possible to 'avoid [imposing] a general analytic distinction between economy and culture on one's material prior to examining its practical constitution' (duGay & Pryke 2002: 12). After all, the insights from this journey make clear how 'cultural economy' is just a different perspective that particularly foregrounds the complexity, multidimensionality and processuality of everyday practices, instead of an actual change in the relation between economy and culture. While Lash and Urry (1994: 64) argue, that it is only recently that the boundaries between economy and culture have become more blurred, the Swahili case serves as an example to show how economically relevant activity has always been cultural.

Furthermore, and maybe even most importantly, this trade journey shows that if we want to understand economic practices in their multiplicity and complexity, we need to get away from a position that takes rationality for granted (cf. Abolafia 1998: 83). When explaining events by reducing them to a fundamental entity or essence, such as rationality or a clear instrumental logic, one can only miss their heterogeneous character. Even though a trader is often defined as an intermediary between producer and consumer, and that is indeed what Ibrahim and Mahir are, they are surely not exclusively economic subjects. This can be well illustrated when looking at the idea of economic rational practice as developed by Bourdieu (Bourdieu 2000: 104). According to Bourdieu, economic rational practice is determined by four dimensions: saving, investing, planning the future and taking risks. In this case, Ibrahim and Mahir hardly do any of this: they borrow money instead of saving it, they do invest but these are only short-time investments, not very planned and calculated but rather influenced by chance or even induced by others, they do not plan far ahead, and the whole journey is rather characterised by spontaneity and flexibility. By relying on their family connections they also do not take big risks. Nevertheless, the whole endeavour is still a way of earning money, just not fitting the often functionalist and rationalist ideal of economic theory. These kinds of trading practices entail disorder, indecisiveness, insecurity and spur-of-the-moment decisions. And it is impossible to decide if it is the aspect of earning a living or the aspect of seeing one's relatives, building bonds and strengthening ties, that is the major driving force of this trade journey.

That the idea of connecting and translocalising through the exchange of material objects is also important to the relatives and friends living on the route as well as for the ones left behind, is expressed in the amount of luggage we carry back to Zanzibar. Although we have sold all the goods that we had brought with us, our luggage is almost as much as when we left. Throughout the journey, we have received numerous calls from our relatives in Zanzibar to bring them things which are either special for the mainland or at least less expensive there than at the coast. For this reason, we now have to carry 5kg of beans, some dried buffalo meet and bottles of honey from Tabora. When we arrive in Dar es Salaam, Aziz, a son of Mzee Farouk from Mpanda, picks us up from the bus station. He and his wife are keen to hear all the news about our trade journey and receive their presents from the interior. Two days later, when we reach Zanzibar, Ruwaida and Rumaitha are eager to get their money. As this journey went very well, Ibrahim and Mahir are able to return the money they had borrowed to buy the goods and they even make a small profit, including the money they spent on the journey.

3ʳᴰ HALT: FINDING ONE'S WAY IN(TO) TRANSLOCAL CONNECTIONS

Although the financial benefit of this trade journey through the Tanzanian hinterland might not be more than as if they would have worked as a shop seller, and sometimes even less, undertaking a trade journey like this one still makes a difference. It is a way of trying to make business, while establishing and maintaining connections over distance and becoming mobile at the same time. Trade and mobile trade journeys in particular are an essential part of what people like Ibrahim and Mahir regard as their Swahili culture and identity and it therefore means a way to take part in this way of life.

As this chapter has shown, belonging to a common translocal space is here as much expressed through practices as it is verbalised. The way in which the trade journey is prepared and organised, the selection and acquisition of goods, as well as the goods themselves tell a lot about the cultural significance of the journey for the traders, but also for the ones left behind in Zanzibar and those visited in the Tanzanian interior. The traders as well as the goods are vital for all parties to actually live and perform their translocal connections. Moreover, it has become clear that these mobile trading practices do not happen in a void but are characterised by the traders' relationships to the places they pass through. Hence, the resiting of imaginative geographies and negotiations between the 'coming from', 'going to' and 'passing through' form an important part of the journey. Engaging with certain topics and debates, such as Swahili history, transport infrastructure, and business opportunities and encountering different experiences and life histories – old and young, mainland and coast, mobility and situatedness – on the route is crucial for constantly (re)creating and (re)negotiating a common sense of belonging. Instead of just 'being around', even on this small economic scale, it is the shared (actual and imagined) experience of undertaking a trade journey that allows the traders to find their way in a Swahili space, which is inherently translocal.

And it is particularly the longing for a stronger position in this space that drives young Swahili like Ibrahim and Mahir into a journey like this one. In general, such a journey has to be seen as a starting point, providing them with experience, connections and reputation. While a journey through the Tanzanian interior is rather simple to organise, as no visa and not many assets are needed, it is regarded as a first step that might allow them more mobility and more economically important trade journeys later. Following the motto *'haba na haba kujaza kibaba'* (generally translated as: a little and a little will finally become a lot) it makes the young traders acquire a taste of translocal lives in their idealised and romantisised form, while at the same time experiencing the effort and struggles that go into translocal connections.

After having given an impression of trade journeys as they are organised by rather inexperienced traders within the national boundaries of Tanzania, in the following chapter I am now going to deal with more professional traders operating on a larger economic scale. Accompanying these traders on their regular business trips to Dubai, following the transportation of goods to Zanzibar and the way they are sold in Zanzibar and Dar es Salaam, I try to give a vivid account of the recent trading connections between the Arabian Peninsula and the East African coast. In doing so, I will particularly engage with the contemporary meaning of these historical links, materially as well as symbolically, pointing at the significance of 'Arabia' as expressed in current Swahili consumption practices. While family connections can also be important for the traders arranging their business trips to Dubai, a clear tendency can be observed to avoid mixing family and business. However, especially when looking at shopping, it becomes clear that also in these more professional translocal connections trade cannot be understood by solely relying on a classical economic perspective, but is instead driven by the cultural values and demands of Swahili consumers.

LIVING (UP TO) THE TRANSLOCAL IMAGINATION: ON AND IN-BETWEEN BUSINESS TRIPS TO DUBAI

'For a coastal society, such as the Swahili of the East African coast, the maritime people of Arabia and the Persian Gulf, or the coastal people of Arabia and the Indonesian archipelago, the sea was not the end of the world, but the beginning of a whole new world of resources and opportunities.' (Sheriff 2006: 15)

Walking from the area of Michenzani through Mchangani into the old city centre of Zanzibar, known as Stone Town, one passes by hundreds of small shops, one next to the other, with most of them selling goods which have been bought either in Southeast Asia (mainly Bangkok), China, India or Dubai. Therefore, these shops are clearly a result of various translocal connections and, while all these places refer to the remaining relevance of trading connections crossing the Indian Ocean, it is the specific link between Zanzibar and Dubai at which I want to take a closer look now, since Dubai and the Arabian Peninsula more widely can be considered as a very important place in translocal Swahili connections. Already in 1988 Constantin and Le Guennec-Coppens point to the enormous role of family connections to Arabia for the constitution of Darajani and Creek Road, the main shopping road in Zanzibar, then also called Dubai Street. Up to today, not much seems to have changed. Spending time in the shops and talking to their owners, Dubai is the place most often referred to when talking about their business trips, either as the place where to do their shopping or as the place to pass through and visit family and friends.

In this chapter, I will thus examine the contemporary mobility between Zanzibar and Dubai and its material effects in the form of shops, by accompanying the everyday practices of the traders involved, on and in-between their business trips. The attraction of Dubai in this context is twofold: on the one hand, Dubai has developed into a famous business paradise for traders from all over the world, on the other hand, the meaning of Dubai for Swahili traders is much more specific. With the historical relationship between Arabia and the East African coast in mind, Dubai today stands for much of the Arab orientation and influence in Swahili culture and identity. In this respect, this chapter focuses on the ways in which the material and imaginative dimension of trading practices that revive these historical links are closely intertwined. Especially from the perspective of consumers, I will show how the traders seem to fulfil a kind of translocal ideal by (re)connecting these two regions and providing them with highly demanded goods, the symbolic meaning of which allows them to express their translocal belonging even without having to be mobile themselves.

BECOMING A BUSINESS TRAVELLER

Traders in Zanzibar are heterogeneous like anywhere else, varying in age, profit, and specialisation (cf. Valcke 1999). Nevertheless, over the last years, there seems to have developed a group of traders, who have all started their businesses about ten years ago when they first travelled to Dubai, now own one or two shops in Zanzibar in which they mainly sell clothes or electronics, and regularly travel to Dubai and further east to buy their goods. Sometimes alone but usually in a small group, they visit the same wholesalers and share the containers for shipment to Zanzibar. Surely not yet being part of the really big business people in Zanzibar, they however form an important part of Zanzibar's private sector with considerable influence on the supply of goods, available fashion and brands. Majid is one of them. At the age of 23 he took the opportunity – together with his friend Suleiman – to enter the UK where both of them worked in a factory in Milton Keynes for a period of three years. Returning to Zanzibar in 1998, the two of them opened a shop in Kiponda, where they first sold original international Music CDs, later focusing on cheaper copies. With the business of CD burning growing rapidly the profit was soon decreasing, so that they decided to open up an Internet-cafe instead. As already mentioned earlier, this Internet-cafe right at the Michenzani roundabout still exists and is now owned by Suleiman alone. Majid however started to focus on clothing, turning their former CD shop in Kiponda into a fashion store. Already in 1999, he travelled to Dubai for the first time, where he stayed with Suleiman's relatives while doing his shopping. Initially flying to Dubai about twice a year, he now travels almost every month to buy goods for his different, expanding and ever changing businesses. Apart from two fashion shops, he has opened a shop selling spare parts for cars and also imports cars. Always looking for new opportunities, and reacting quickly when business opportunities change, his mobility between Zanzibar and Dubai has become a central part of his everyday life. Being a frequent traveller with an Emirates gold card, he often gets price reductions and sometimes even a free flight. One time, Majid and I met when boarding the aircraft to Dubai at Julius Nyerere Airport in Dar es Salaam. Both of us were supposed to stay with Rashid, one of Suleiman's cousins, who would come to pick us up at the airport.

Fig. 18: Emirates' advertisement in Dar es Salaam

DUBAI: BUSINESS WITH AND WITHOUT FAMILY

Although I had met Rashid once before when he was visiting Zanzibar, I hardly recognise him at the airport. While he had been wearing Blue Jeans and a shirt in Zanzibar, he now wears the classical long white dress (*kanzu*) and a turban (*kilemba*), as most Emirati men do. On the other hand, when entering his house in Jumairah, it feels like entering a house in Zanzibar. The furniture, the decoration, the clothes: everything could be found in Zanzibar as well. But, instead of seeing this house in Dubai as a copy of its version in Zanzibar, it rather indicates how close the connections are between these two places, with many of the things in Zanzibar actually being imported from Dubai, and a lot of the others taken from Zanzibar.

Rashid was born in Kikwajuni, an area of Zanzibar Town in 1952. Considered as Arabs in the population statistics of that time, his family suffered a lot from the aggression in the context of the so-called Zanzibar Revolution, which took place in January 1964. In its aftermath Rashid himself had been imprisoned for almost two years. Finally, in 1972, he manages to leave the island and move to Dar es Salaam, where he started some trading activities. Two years later, in 1974, he followed his older brother Khamis to Dubai, who had already gone there in 1968. In Dubai, Rashid first joined the army, as it offered him a secure position in his status as a refugee. In the meantime, he has done a couple of other jobs, mainly in the import-export sector, and now uses his knowledge and connections to help friends and relatives like Majid with their trading businesses. Watching these two men on their daily tours through the shopping areas and listening to their reports when returning home reveals a great deal about how these professional trading practices are closely related to the careers of former Zanzibar refugees.

Business with family

When Majid is in Zanzibar, Rashid keeps Majid informed about the schedule for the container shipment to Dar es Salaam and Zanzibar, so that Majid can arrange his deals at the right time, trying to keep the timeframe between buying and selling as short as possible. Moreover, it is Rashid's job to observe the market in order to make Majid's relatively short stays – sometimes only lasting two or three days – as efficient as possible. Early the next morning, Majid and Rashid leave the house to go on their first exploration tour.

To refill his shops, Rashid guides Majid through the big wholesale areas, hidden to most of the tourists, where rather cheap imports mostly from Asia can be obtained. On the one hand, only these allow for a considerable profit being affordable for a larger number of consumers in Zanzibar, on the other hand, they enable traders with smaller assets to engage in trading practices as well. Nevertheless, also at these wholesalers goods of different qualities and prices can be bought, and it is important to Majid to get good offers for products of middling or even relatively high quality as this also represents his status as a trader. He wants to stand out as having a shop known for good and fashionable clothes as well as he wants to be regarded as a reliable spare parts and car trader.

Since Rashid has already visited some of the car dealers before to look at the cars on sale, he can now direct Majid's attention to the most promising offers. But, even though Majid and Rashid constantly reflect on what cars are suitable for – and wanted in – Tanzania, it remains hard to estimate the profit. As Majid states, whereas one car has brought him a profit of 1 400 000 Tanzanian Shilling (TSh), at that time approximating 700–800 €, it has also happened that he was forced to finally sell a car without any profit on his side at all. With regard to clothes and spare parts, it is easier to predict the anticipated turnover. In general, traders aim to sell a good at about two to three times the wholesale price in order to cover their costs and make the business worthwhile. When they cannot expect to find enough people to buy their goods at that price, they rather do not acquire it at all. In this respect, it is crucial for Majid to know the market in Zanzibar, in order to be able to estimate the amount people would be willing to pay for a certain good, while Rashid helps him to find these goods at the lowest available price. Over the last years, Rashid and Majid have done a lot of business together. If Rashid has some capital, he sometimes acquires a share of the goods or buys some on his own, which Majid would take in his container and sell for him. This shows that these translocal connections are not one-sided, not only organised and fostered by people mainly living along the East African coast, but are, to a big extent, also triggered and maintained through Swahili people residing elsewhere. Hence, it is their points of view, their practices and imaginations that also have to be taken into account when trying to get to grips with the constitution of the translocal Swahili space.

A sidenote on the situation of Swahili in the United Arab Emirates

An important moment for the movement of Swahili to the United Arab Emirates (UAE) was the Zanzibar Revolution in 1964. This is also the context in which Rashid and his brothers have moved to Dubai. What started as a one-directional movement between Zanzibar and the UAE as a result of the Zanzibar Revolution in 1964 has however been followed by a more general mobility of Swahili from other places along the East African coast, as well as from places in Europe and the United States. Family reunions, movements to and away from the UAE have led to a fluctuating Swahili population in the UAE that overall still seems to be increasing. Yet, it is hard to tell how many Swahili are currently residing in the United Arab Emirates (UAE), since they are not listed as a separate category and many of them have meanwhile been fully included as Emirati nationals in the official population statistics and censuses. The Swahili, who have more recently moved to the UAE with a Kenyan, Tanzanian or for example British Passport, are not distinguishable in the statistics from other citizens of these countries and thus cannot be clearly recognised as Swahili either.

Although the UAE have meanwhile become closely enmeshed in translocal Swahili connections to various places, it is clearly the Zanzibar Revolution that has to be considered as the main trigger of this movement. In the aftermath of the political upheaval in January 1964, residents that were considered as Arabs were deported, the last Sultan Sayyid Jamshid bin Abdallah bin Khalifa bin Harib fled to Portsmouth in the UK, and many were killed. Whereas estimates say that about 6000–10 000 have died (cf. Cameron 2004: 105, Clayton 1981, Gilbert 2007: 170), it is assumed that about 30 000 people of Arab origin were forcibly expelled or fled at their own initiative, many of them heading towards Arabia (cf. Bakari 2001: 193).

Though Zanzibar had long been part of the Sultanate of Oman and many of the refugees could clearly trace their relatives in this country, the Omani government under the rule of Sultan Saeed bin Taymur (1932–1970) refused to organise a collective repatriation process and, through the end of 1964, only agreed on accommodating 3700 Zanzibari (Peterson 2004: 46). Apart from a general reluctance of the Sultan of Oman until 1970, this was also reasoned by the fact that people who could be identified as former opponents or critics of the Sultan were denied access. Furthermore, Oman only accepted Zanzibari refugees who could prove to belong to one of the Omani tribes (*kabila*) (cf. UNHCR Archive Box 15/112 01–1964 Evacuation of Manga Arabs from Zanzibar File 94, 17.08.1964). Only when Sultan Qaboos overthrew his father in July 1970, some of the Swahili who had first settled in other Arabian countries such as the United Arab Emirates (then still the Trucial Sheikhdoms), Kuwait, Egypt and Saudi Arabia or in Europe, were able to move to Oman as they were invited to join forces and contribute to the 'awakening of the country' (Valeri 2007: 485). Townsend (1977), for example, estimated that by the end of 1975 between 8000 and 10 000 people from Zanzibar had settled in Oman, which are assumed to have increased until today, now consisting of 100 000 (Valeri 2007: 486). According to Al Rasheed (2005: 100) Zanzibari themselves claim to be even more, suggesting numbers of 300 000 already at the end of the nineties, due to natural population growth, and more recent, often economically motivated migration, accompanied by a wave of Omani migrants from Burundi and Rwanda at the beginning of the 1990es, which are also often termed Zanzibari. Overall, Swahili have definitely come to form a considerable community in Oman, which has played and is still playing a crucial role in the process of nation building and the creation and negotiation of an Omani identity.

On the contrary, the Sheikh of Dubai, Rasheed bin Saeed Al Maktoum (1958–1990), was more hospitable and generally accepted Zanzibari refugees from 1964 onwards. So, even when Zanzibari were finally allowed to enter Oman in bigger numbers, many of those who had arrived in Dubai before 1970, decided to stay and have since motivated other family members to follow them. Soon after the revolution and still in the 1970es, it can be estimated that even more Zanzibari refugees were residing in Dubai than in Oman. Nevertheless, they have never played an important role in national debates. Although the United Arab Emirates today host a large number of Swahili, they have hardly been visible as a distinct group in public discourses.

It is only over the last couple of years, that academics have started to engage with the situation of Swahili in Arabia. However, most of the literature so far, has mainly concentrated on Zanzibari in Oman (al Rasheed 2005, Valerie 2007, Peterson 2004) or Washihiri and Wahadimu in Yemen (Walker 2008, 2010), tracing the ways of families who migrated to East Africa back to their former home and looking particularly at the ways in which these historical relations are negotiated in the national contexts as well as in individual identities today. Al Rasheed, for example, traces the life histories and experiences of those Zanzibari who returned to Oman in the 1970es putting a special emphasis on the transnational context of what has become a Muscat elite (Al Rasheed 2005). On the other hand, Peterson and Valeri rather focus on the role of Zanzibari for the Oman nation-state, looking at their group feeling (asabiyya) and identity as a valuable contribution to the study of modern state-led processes of Oman nation building (Valerie 2007, Peterson 2004). Despite the different contexts (transnational *vs* national), what is commonly agreed upon is that the Zanzibari refugees were welcome participants especially in the early states of Oman's development after 1970, not only because they were educated and possessed skills that were in high demand, such as the knowledge of English, but also as they were considered as insiders of Oman on grounds of their ancestry while at the same time being outsiders to the tribal alliances of rivalries due to their long absence (Al Rasheed 2005: 102, Valeri 2007: 485, Peterson 2004: 46). Nevertheless, whereas the Zanzibari in Oman were on the one hand appreciated for their education and their proficiency in English, on the other hand, they had to face negative attitudes and refusal due to their 'Africanness', most clearly recognisable in the Swahili language and a weakness in Arabic which some remain until today, and their more liberal and 'cosmopolitan' traits (Al Rasheed 2005: 103, Valeri 2007: 489).

In the UAE, the Swahili were able to remain more inconspicuous than in Oman, as the establishment of the oil industry and the rapid economic growth brought along with it an acceleration of immigration from different places all over the world, with other groups from India and Pakistan clearly outnumbering and driving the attention away from them. In the UAE, Swahili have long been less integrated into the Emirati society and, regarding the integration in the job market they have hardly reached any high positions. Whereas some have initially entered the army, others soon started to establish their own businesses, mostly consisting of shipping goods to East Africa (Thani 2003: 241). As only UAE national are allowed to own a shop, they often had to rely on businesses on-the-go, in-between wholesalers, agents and the harbour. Only their children, born in the United Arab Emirates, fluent in Arabic and with better education, are now able to acquire good and more stable positions in the job market. On the contrary, Swahili in Oman have soon been strongly represented in the national oil company (Petroleum Development Oman), as well as in the Ministry of Defence, the Omani Intelligence Agency and the police (cf. Al Rasheed 2005: 102–103, Valeri 2007: 494, Peterson

2004: 47–51). Moreover, Swahili have developed a strong position in the tourist sector in both countries.

Instead of either being celebrated or critically discussed as 'homecomers', in Dubai, the Swahili rather seem to be one migrant-group of many. Here, however, contrary to the situation in Oman, Zanzibari refugees were allowed to found their own association (cf. Bakari 2001: 192). This Zanzibar Association was mainly occupied with the rights of Zanzibari refugees. Moreover, many of its members have been emotionally involved in Zanzibari politics, some of them even actively. However, different political orientations and opinions have led to disagreements and conflicts and contributed considerably to the dissolving of the political part of the Zanzibar Association in the 1990s. Besides, with most Zanzibari by now having gained Emirati citizenship, the original aim of the association to secure their rights has been accomplished. One crucial aspect, concerning their rights, was to secure adequate housing. Due to their calls to the government, Sheikh Rasheed bin Saeed al Maktoum, with the support of the UNHCR, gave 101 houses to Zanzibari refugees in Al Rashidya (Thani 2003: 245), a residential area near the airport, and built a number of apartment blocks to accommodate 8000 people in Al Karama (cf. Elsheshtawi 2008: 983). Al Rashidiya is one of the relatively wealthy neighbourhoods of Dubai with a large Arabic population and also favoured by European expatriates. As 101 houses have been given to Zanzibari refugees, even though some have since moved to other places, the number of Zanzibari in the area is still high. The mosque in that area is known as the 'Zanzibari Mosque' and this is also were the Association still owns some space, where communal functions such as weddings are held.

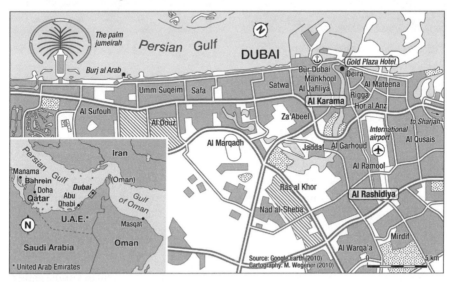

Fig. 19: Map of Dubai

Concerning the residential integration in Oman, especially in Muscat and Matrah, in the 1970s and 80s, some quarters of the cities were known for a high percentage of Zanzibari residents (e.g. Wadi Adey). Today, this has become a lot less clear, though famous places named after locations in Zanzibar such as Mnazimmoja and Khor Pemba still expressively signify the entangled history. The topicality of these relations in Oman is also vividly expressed by the recent book by Al-Riyami (2009), a Zanzibari Omani himself, in which he presents his view on the history of Zanzibar and Oman, being a bestseller and sold out shortly after its publication. Whereas Zanzibari in Oman and the historical relationship between these two places have remained an intensely discussed topic of national concern, fuelled by books like this one, but also in form of groups in online networks such as Facebook and the like, it seems that Swahili residing in the United Arab Emirates rather deal with this subject on an individual or family basis. Coming back to Rashid and his family and their involvement in the trading connections between Dubai and Zanzibar, it becomes clear how their situation mirrors this briefly sketched general history of Swahili in the UAE.

Fig. 20: Family meetings in Dubai

In the evening, when Rashid and Majid return from their tour through the shops we all meet for a big family meal in the house of Rashid's brother Khamis in Al Rashidiya. Khamis is the third of ten brothers and sisters and the oldest son. He was the first one in the family to move to the United Arab Emirates in 1968, where he soon became a member of the Zanzibar Association and was later even appointed as their General Secretary. It was while fulfilling this function that he got allocated one of the houses built for the Zanzibari refugees in Al Rashidiya. In the meantime, many of his closer relatives have followed him to the United Arab Emirates, so that most of the brothers and sisters get to see each other regularly. While this suggests a certain stability, looking closer at the individuals however reveals the ongoing mobility of most of them, between the United Arab Emirates and East Africa, but also between the UAE and Oman, the UAE and the UK as well as to places in Asia (esp. India and Malaysia). This particular family gathering for example consists of 'Zanzibari refugees', Emirati business men and newly-appointed doctors and nurses, traders and family guests from Zanzibar and Mombasa and their Emirati-born cousins – all, apart from me, belonging to the same

Arabic *kabila* – highlighting the multiple ties and longstanding mobility between places in East Africa and Arabia.

In academic literature, relations between the place of current residence and the place one had to leave are often regarded through the lens of nostalgia, assumed to be characterised by the longing for a lost or imagined past (cf. Fortier 1999, Nora 1989, Shaw & Chase 1989). Also in respect to Zanzibari living elsewhere and their images of Zanzibar, Swahili scholars like Sheriff have referred to a 'sometimes very nostalgic image of Zanzibar as "the island of dreams"' (Sheriff 2008: 78). Topan (2006), as well, refers to the sentimental image of an imagined Swahili community in Zanzibar. Whereas I do not want to deny that some representations of Zanzibar indeed consist of an idealised narrative of the two islands, what becomes clear in the case of Rashid and his family is that, to most of them, these relations are a lot more active and pragmatic. With the majority of them having received UAE passports, there is no problem to travel to either Kenya or Tanzania apart from financial constraints. Hence, Rashid, his brothers and sisters, and their families now regularly go to Zanzibar to attend weddings, make visits to severly sick relatives, join a mourning or simply spend their summer holiday there to escape the heat on the Arabian peninsula and see their relatives. When their brother Salman died, Rashid married his widow as a second wife and therefore spends almost half of the year in Zanzibar to stay with her. When Abdullah's daughter Salma wanted to study medicine at university, she was sent to Dar es Salaam where one of her grandmothers still lives, since there she was at that time still eligible to obtain a national scholarship. This shows that, on the one hand, it is clearly the relatedness that makes these people travel between the two places in order to maintain, strengthen and simply live the translocal family. On the other hand, especially when it comes to Khamis, Rashid and their brothers, their relations are also strongly characterised by trading activities.

Khamis for example often joins Majid and Rashid on their tours to the different car dealers, as he also ships cars to Mombasa, where his brother-in-law sells them for him. Their brother Abdullah, very soon after his arrival in Dubai, started to trade in spare parts, and it seems as if all of them gain considerable amounts of their overall income out of their small-scale trade with East Africa. Nevertheless, while these trading activities still play an important economic role to them, most of their children have been to Arabic or English-speaking schools, and now either still continue their studies at the university or have already entered the national job market. None of their sons has their knowledge of where to get good offers, how to arrange the shipments and what to consider when buying the goods. When Majid comes to Dubai, he therefore takes Rashid as his guide, makes a business deal with Abdullah or shares a container with Khamis and his brother-in-law in Mombasa.

Overall, joining Majid on his business trip to Dubai shows how the motivations and roles of the people involved differ. As the imaginative geographies and senses of home vary between family members, continuously negotiating the inbetweeness between Arabia and Africa under the influence of changing political and social contexts, so do their practices. Whereas for some the economic use, or at least the attempt to financially benefit from these relations, is crucial and has been part of their lives almost since their arrival in Dubai, others – and especially the younger generation – have turned their attention more to the national job market, where they are welcomed and enjoy the benefits of the Emirati citizens. While for several hundred years, Arabs have moved to Africa hoping to improve their lives by taking part in trading activities and the more prosperous life in Zanzibar, along the East African coast, or even in the interior, today it mainly, but not exclusively, seems to have turned the other way around. But, although the economic and political contexts, and with it, the direction and intensity of the flows of people, ideas and goods have changed, the relations have still remained their power in terms of processes of identification as well as with regard to providing a livelihood. The way in which Majid organises his business trips therefore points to the complexity and flexibility of these connections, their multiple dimensions, their processual and multidirectional nature. As Majid's trip emphasises, his business is definitely the reason for his mobility. Different to the trade journey through the Tanzanian hinterland, in the case of Dubai, business trips can be more easily separated from family motivated mobility. Nevertheless, these trips are still very closely interwoven with and facilitated by family relations, and in the case of Majid and Rashid often even merge into a joint endeavour.

Business without family: The Gold Plaza Hotel

Not all of the traders who travel to Dubai to buy the goods they want to sell in their shops see it as an advantage to involve their relatives in their trading activities. Khadija is one of those. She lives in Michenzani, Zanzibar Town, with her two children, while her husband is away for most of the time working on container vessels in the Mediterranean. Khadija herself had been working in a bank, but about two years ago when she lost her job she decided to open a shop instead. In one of the smaller spaces in the containers of Michenzani, she sells women's and children's clothes, long wide women's dresses (*madira*), fashion jewellery, perfume and headscarfs (*mitandio*). This is already the third time she travels to Dubai, and although she also has relatives living there, she prefers to manage her business activities without them. Feeling that, staying with them, she would only loose valuable time while having to join in the family activities and would also have to spend more money when having to take a taxi to reach the shops from their location, she chooses to stay at the Gold Plaza, a hotel right next to the Gold Souk of Dubai in the middle of numerous wholesalers.

The Gold Plaza Hotel is owned by an Indian family, and – at least for the last ten years – it has become famous for its Tanzanian and mainly Zanzibari guests. Even in the travel guides it is mentioned that the hotel mainly hosts East African traders (Carter & Dunston 2007: 148, Gilmore 2007: 101). When entering the hotel, already at the reception two men sit there discussing their latest acquisitions in Kiswahili. In the corridor of the first floor, next to a weighing scale, there are a number of big boxes addressed to Dar es Salaam and Khadija immediately recognises some of the names written on the boxes. Usually, she travels with a female friend who also owns a shop in Zanzibar so they could share a room and therefore reduce the costs. Moreover, it is easier to do the shopping together as they can give each other advise on what to buy.

The two women leave the hotel early in the morning to arrive at the shops as soon as they open. Slowly they move from shop to shop looking at the goods on sale, asking for their wholesale price and taking their business card if the goods are of interest to them. When they have got an overview about the variety of products and good offers, they return to the selected shops on the following day to buy the chosen goods and have them directly send to either their cargo agent or at least to the hotel. One day they reserve for the shops where they have bought a lot last time and expect to do so this time as well. These three to four days that are needed to complete the shopping, they stay in the Gold Plaza Hotel, which is perfectly situated in the centre of their trading activities. Afterwards, the two women separate and move to their respective relatives to spend a couple of days among them. Relaxed, with their goods already placed at the agent's, Khadija can then enjoy the company of her family members, completely giving in to their plans and arrangements.

Fig. 21: Gold Plaza Hotel in Dubai

Massoud, another trader from Zanzibar who owns two shops in Mchangani, names similar reasons for staying at the Gold Plaza Hotel when being in Dubai. Whereas most of his relatives do not live in the city centre, it would be more complicated to move between their houses and the wholesalers. Moreover, they hardly have the space to store his luggage, whereas the Gold Plaza has all the facilities needed and can easily arrange the transport of his boxes to the harbour. Besides, most of his closer relatives whom he would not mind to ask helping him on his tours from shop to shop live in Oman. Not having very close connections to those in Dubai, he considers it inappropriate to bother them, admitting that, with his business in mind, he would not be a very agreeable guest. Often only staying for two or tree days, there would not be much time left to spend with his relatives.

These two examples illustrate how business trips to Dubai can also be arranged alternatively, without involving any relatives in the actual trading activities. Compared to Majid, who feels to benefit from the knowledge, interest and active participation of Rashid and his brothers, traders like Khadija and Massoud rather see their relatives as a disturbance in respect to their business trips. They have either decided to completely separate family from business and thus arrange these two successively, or they even do not visit their relatives at all. Assuming that their busy schedule would not meet their relatives understanding, they prefer to be completely independent. On the contrary, when Majid and Rashid leave the house early in the morning and only return late in the evening, they know that dinner will most probably still be waiting for them. Rashid's wife is already used to their unpredictability, and when the two traders call in advance, they can be sure that food will be prepared for them. While Massoud mainly eats on the go, just briefly stopping at one of the numerous fast food places in town, Khadija and her friend feel rather uncomfortable eating in public. Often, they prefer to buy something in one of the cafes near the hotel and take it to their rooms, Moreover, they like to return to the hotel in the middle of the day to relax but also to meet some of the other guests and exchange news and information about good offers – an opportunity they would not have when staying with family or friends.

Putting business before family

Overall, when comparing these business trips to Dubai to the trade journey through the Tanzanian hinterland, it is striking that, even in the cases in which family members become involved in the trading business, the stays in Dubai are much more subject to economic calculation. The frequent trips to Dubai are first of all business endeavours that are organised in a way to secure the highest possible profit. Whereas in some cases family members are regarded as a valuable partner to increase the profit, helping to find good offers or reducing the cost of the stay by providing free accommodation and food, in other cases family is rather avoided as it is perceived as a hindrance to the busy schedule of the traders. This evaluation mainly depends on the knowledge and understanding of the relatives concerning the trading business.

From the perspective of the relatives involved in the trading practices, it is also mainly the economic aspects that determine the course of the stay of the Swahili traders in Dubai. Only when shops are closed, the time is used to spend time with the family or do some sightseeing. This corresponds to the fact, that for these relatives engaging in this kind of translocal trading activities forms a central part of their lives as well. Whereas most of the relatives visited on the trade journey through the interior of Tanzania have businesses that run independent from their relatives at the coast, in this case, the traders residing in Dubai rely on the cooperation with their family members in Zanzibar in order to succeed with their own affairs.

Although on business trips to Dubai the meeting of family members thus seems to be reduced to a minor matter, it still essentially contributes to the enlivening of family bonds between the UAE and the East African coast, as it involves regular communication, the exchange of family news and a more general sense of involvement in each others lives. Due to the high frequency of business trips to Dubai, some of the traders even become considered as being a part of their (host) families in the UAE, with their recurring presences turning into a common and very mundane phenomenon.

However, the business trip itself is only one dimension of the life of a trader engaging in these kinds of trading practices. Despite their high mobility between Zanzibar and Dubai, being a trader on this scale also involves more or less long periods of situatedness. Historically, these trading connections across the Indian Ocean have been characterised in particular by the long periods of sailing from one place to the next. While the traders themselves now enter an airplane that takes them from Dar es Salaam to Dubai in less than six hours, it is their goods that still take weeks and sometimes even months to arrive in Zanzibar. After each trip the traders thus first have to face a rather passive and immobile phase of waiting for the goods, which are still subject to the dynamics of Indian Ocean travel. Though, on this route, traditional dhows have meanwhile given way to large containerships, the tension that was involved when awaiting the fleet of dhows with new cargo that dominated the trader's life many decades ago, is still present today. Also today a delay or even loss of goods has severe economic effects and thus threatens the traders' existence.

THE TRANSPORTATION OF GOODS ACROSS THE INDIAN OCEAN

'Shadowy cargo vessels cheerlessly ply these waters, bringers of unaffordable goods, reminders of deprivation, enticements to get up and go. Silent pipers, whom we follow by jet planes, those who can, and stretch ourselves between lives as contrary as the ends of the cross.' (Vassanji 1991: 130)

As Sheriff (2008: 62) points out, with regard to the relations between Arabia and Africa, 'the sea has always been the main medium of transport, exchange and intercourse'. Generally, the history of material exchange crossing the Indian Ocean has been closely connected to both the dhow and the monsoon. While the dhow trade involved a large 'floating population' giving rise to what has been called a 'maritime ethos' (Sheriff 2006: 29), the monsoon winds were responsible for a clear timeframe of the trade, on the one hand facilitating seasonal movements in one direction by a rather steady breeze from the back, but on the other hand, also restricting this movement and making it necessary for traders to stay in the harbour and wait until the direction of the wind has changed before embarking on the return journey. Nevertheless, with an estimate of eighty dhows a year between Arabia and the East African coast from 1907–1947 (de Vere Allen 1993: 241, Villiers 1952), the dhow has long been crucial for the transportation of goods, people and ideas across the Indian Ocean.

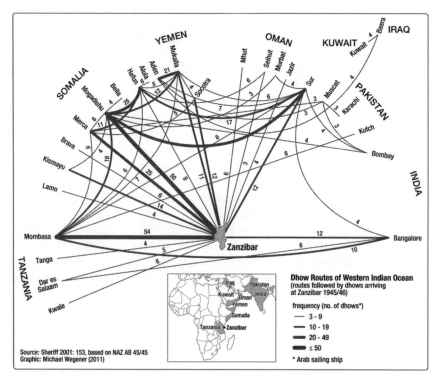

Fig. 22: Number and direction of dhows in the Western Indian Ocean 1946

Concerning the material exchange, the dhow trade suffered from the competition of the big American and European sailing vessels and especially the steamship, which soon took over the transport of the valuable cargoes such as ivory, cloves and mangrove wood. By the late 1950s, dhows were only left to carry a tenth of the total trade of Zanzibar (Sheriff 2008: 67). However, concentrating on bulkier commodities and less accessible ports and routes, they have yet remained to play an important role in the maritime trade along the East African coast (cf. Prins 1965: 5). Moreover, dhows are still used as one of the main symbols to represent Swahili culture, often evoking a rather romantic image of Swahili history.

Whereas the image of the dhow is used to refer to the maritime character and particularly the outward oriented and cosmopolitan worldview often assigned to the Swahili, as well as to remind of the economic power and more prosperous past of Zanzibar, the large cargo ships now dominating the maritime trade are rather incorporated into the wider discourses of a critique of capitalism and western dominance. This becomes especially clear when reading Villiers, who himself travelled across the Indian Ocean in a dhow:

> 'Now no ship sails upright and graceful, westwards upon the breath of the Indian Ocean trades. No ship storms eastwards before the wild west winds, under close-reefed sails [...].

Such ships [dhows] still understand the way of the Indian ocean and its winds and weather, but for some years now all European vessels, which carry the bulk of the vast commerce there, have been driven by power.' (Villiers 1952: 13)

Although indeed most Swahili seem to have very positive associations concerning the dhow, listening to the accounts of the traders reveals a much more neutral and pragmatic opinion in respect to the new mode of transport than the ones observed in the academic literature. Today, all traders importing goods from Dubai have to rely on the large container vessels, regularly moving between Dubai and the East African coast. While from a more distant perspective container shipment generally epitomises a very regulated, ordered and scheduled space, from the perspective of the traders the transportation of the goods comes with a number of practical concerns that reflect the irregularities and disorder they often have to face when waiting for their goods. This phase, in which their new merchandise is on its way from Dubai to Zanzibar, is thus generally characterised by a certain tension and insecurity.

Leaving Dubai about twice a month – the exact date usually depends on when the ship is fully loaded – some containers still need four months until they can be picked up in Zanzibar. Only every two months there is a ship that follows a more direct route decreasing the time of shipment to two to three weeks. That means that all their business trips have to be well scheduled in order not to miss the departure of the vessel and, that all business plans have to be made long in advance so that the goods arrive on time avoiding an empty shop. Although the traders try to take as much as possible in their regular luggage on the airplane, as soon as their businesses have reached a certain size, they cannot avoid the container economy, since arranging the transport by air would be far too expensive. That means that after a trip they all face up to four months of waiting for the goods to arrive even if no complications occur.

Most traders work with cargo agents who are supposed to take care of the luggage from the harbour in Dubai to the clearing in the harbour of Zanzibar. While the phase of its shipment is relatively unproblematic, it always becomes more critical when the container arrives at the harbour. According to estimates of the Vero Marine Insurance Limited every year approximately 2000 of 13.6 Million containers get lost on their way through the world (Bogatu 2007: 89–90). And, indeed, many traders mention that on the route along the East African coast containers regularly fall into the water due to very bad packing and stapling and sometimes they simply disappear in one of the harbours along the route. The bigger vessels that take the long route along the coast all stop in Mombasa and Dar es Salaam where the containers have to be loaded on to a smaller vessel to be taken to Zanzibar. Especially in Dar es Salaam delays occur regularly. Due to the narrow entrance to the harbour, it is hardly possible to have more than five ships entering and five ships leaving the harbour in a day. When the container has finally arrived in Zanzibar, in the worst case, it might still take two to three months until a container will be released. This phase, when the goods are in the hands of the custom, is especially annoying to the traders. Apparently, it is not uncommon that containers are broken open and goods are stolen. As a result, the time of waiting for the goods is usually pervaded by a constant worry about the shipment, and the traders are only relieved when they are finally allowed to pick up their goods from the harbour. Until then, they sit in their shops or pass by other traders hoping to hear news about the momentary location of the vessel, ask the agent for his information, or go directly to the harbour trying to catch a glimpse of their luggage.

A central aspect with regard to the planning of business trips to Dubai is to calculate the arrival of the goods before the beginning of Ramadhan. In Zanzibar, this month is clearly the most important time of the year in respect to trading activities. This is when the sales rates denote an explosive boost as most people do their shopping for the *Eid-il-fitr* celebrations at the end of the month. In order to be well prepared for the rush of customers, traders usually aim at receiving their goods at least one month before the beginning of Ramadhan. However, it regularly occurs that containers are delayed or that, even if the container has already arrived in Zanzibar, there is no way to access the goods on time as they are kept at the customs. Realising that they would otherwise completely miss out the peak of trading activities, something that would extremely reduce the gains of the business for the whole year, many traders are forced to travel to Dubai again to buy as many goods as they can while remaining within their 40kg luggage allowances on the airplane. This way, they at least have some new goods to fill their shops with – and make it more attractive to the customers. Sometimes even those, whose container has arrived on time, still travel to Dubai during the month of Ramadhan in order to quickly refill their shelves and try to make the most of this precious time.

Thus, although the traders themselves now travel by airplane, it is still the maritime trade that strongly influences the time frame of their businesses. Despite moving independent from the monsoon, especially due to a weak infrastructure in the harbour, the journey of the container in some cases still takes up to 4 months and is accompanied by a number of risks. The connections have changed, with large container vessels now leaving from Dubai and not from the old dhow harbour of Sur in Oman; the directions of the flows have changed with more goods now travelling from Dubai to Africa than the other way around; and the goods themselves have changed from spices, ivory and mangrove wood to clothes, electronics and cars. Nevertheless, it is still the maritime trade that sets the rhythm of many of the contemporary business activities in Zanzibar and along the East African coast, with the transportation of goods by airplane only forming a small addition. Also in this form of container economy, traders still anxiously wait for their goods, hoping for them to be complete and on time, tensely following the mobility and immobility of the vessels on their way across the Indian Ocean.

KEEPING SHOPS

As soon as the goods can be picked up in the harbour, they are taken to the shops of the traders. Trading in this quantity makes it necessary for these traders, unlike the ones travelling through the Tanzanian interior, to provide fixed spaces where they can store and offer the goods to the customers. In effect, apart from regularly travelling to Dubai to acquire new merchandise, a big part of their job consists of looking after and keeping their shops. Based on the experiences of Majid, Khadija and Massoud, who I have accompanied on their business trips to Dubai, in the following, I will thus give a more detailed impression of how traders organise this part of their trading practices. Decisions have to be made where to rent a space, always negotiating bearable costs and favourable spots, or whom to employ as a shop assistant. Finally, as long as the trading business goes well, there is also the opportunity to expand and open a new shop somewhere else, preferably in the booming shopping areas of Dar es Salaam. Whereas certain shopping areas in Zanzibar and Dar es Salaam can clearly be identified as material manifestations of the translocal connections between the East African coast, Dubai, and other places across the Indian Ocean such as Bangkok and Singapore, this section attempts to go beyond a simple mapping of these spaces, by particularly pointing to the complexities and instabilities involved in keeping shops.

Retail geography is often recognised as having long been 'a largely descriptive and all too often simplistic mapping of store location, location, location' (Crewe 2000: 275). Nevertheless, what has formerly been assigned to be one of the 'most boring of fields' in geography (Blomley 1996) has in recent years extremely strengthened its position in the social sciences and humanities more generally. As Crewe illustrates in respect to geography, turning away from the strong focus on marketing and mapping, many retail geographers now attempt to follow an ethnographic approach, striving for a more complex and nuanced understanding of the social and cultural significance of contemporary consumption practices (Crewe 2000: 284). This has especially broadened the perspective with regard to the everyday and rather mundane aspects involved in retail activities. Also in my case, hanging out in the shops, helping to arrange newly arrived goods, serving customers or just waiting for the next one to come in gave me a very valuable insight into the everyday character of the trading business.

Kiponda and Macontainer

Kiponda is one of the busiest shopping areas in Zanzibar consisting of narrow streets stretching from the market into the interior of Stone Town. It is an old quarter – and the only one in which already in 1893 stone houses outnumbered huts – that was mainly inhabited by Indian and Arab merchants (cf. Sheriff 1995: 25–26). In the two to often three storey houses, the space downstairs was usually reserved for the shop in front and a store in the back. Upstairs were the living rooms, and generally a balcony allowed for a view over the narrow streets. While this usually meant that the owner of the house is also the owner of the shop, nowadays some also rent out their shops, thus enabling traders living in other parts of the town to have a business there. Over the last years, the number of shops selling the latest fashion imported from Dubai and Asia has increased steadily and many have started to concentrate on the rather high priced sector of retail activities in Zanzibar. Women's clothes, gold shining fashion jewelleries, creams, make-up and lotion, glittery adorned materials, long black *mabuibui* and colourful heafscarfs (*mitandio*), sporadically also Punjabi dresses, and a great selection of female shoes fill most of the display windows of Kiponda today. This is also the area in which Majid has the shop that has been the most stable location involved in his trading activities. It is his shop and the one of his friend situated just around the corner, that have specialised in selling boys' and men's clothes, otherwise offered on a bigger scale in Mchangani, on the other side of the Creek Road.

Fig. 23: Kiponda

Fig. 24: Selling boys' clothes in Kiponda

Majid hardly stays in this shop for more than half an hour at a time before leaving again to follow one of his other businesses or to look after his other shop in Mkunazini in which he sells spare parts for cars. He usually drops in a couple of times a day to say hello and ask how things are going and exchanges some news with his shop assistant. When he has some more time, he visits his friend in his shop around the corner. There, they sit and talk, often joined by some other friends and neighbours, plan the next trip, listen to music or simply watch people passing by. When there is not much business to do, they sometimes play domino or draughts to pass time. Massoud, as well, spends most of the time on the bench (*baraza*) in front of his electronics shop in Mchangani. Only in the shop in which he sells clothes he employs a shop assistant and generally manages the sale of the electronics himself.

The area called Macontainer, where Khadija has her shop, especially represents the female trading sphere. Behind the high apartment blocks of Michenzani, since 1995, when the land was made available by the government, containers have been placed in rows, forming one of the most frequented shopping areas for women, who mainly come from the area but also from Stone Town or further away. Apart from a few shops selling hardware and some things for interior design, most of the shops have concentrated on female clothing and shoes, offering a wide selection of the latest fashion mainly acquired in Dubai. Furthermore, some shops selling baby and children's clothes, as well as a larger store exclusively selling all kinds of diapers meet the demands of mothers. Some beauty salons and shops selling cosmetics seem to perfect the range.

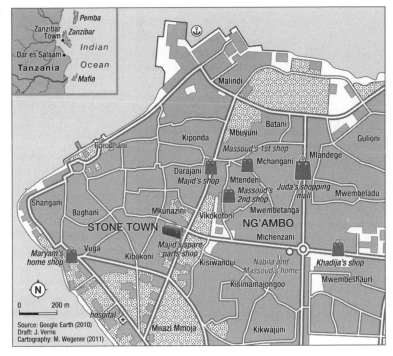

Fig. 25: Map of Zanzibar town's main shopping areas

Khadija deliberately chose this area to open her new shop, as it is close to her flat in Michenzani, so that she can easily go home during the day, for example to pray or to eat. Moreover, Macontainer is very popular among the women she knows, and they all pass by her shop to look at her selection when they go shopping there. And with so many shops within such a small distance, she easily got to know many of the other traders so that she can exchange her ideas with them, discuss what is in fashion or plan the next trip together. Therefore, although being a rather nondescript area from afar, by most people in Zanzibar Macontainer is definitely recognised as a shopping hub and a clear and visible expression of particularly female translocal mobility.

When Khadija cannot be in her shop herself, she usually asks one of her younger relatives to look after it. She just briefly tells them the prices and reminds them to write down everything they sell in her absence. Sitting in the shops, one quickly finds out, that bargaining does not play a big role in this kind of trade. When placing new merchandise in the shop, the goods are usually ordered by price. They are all assigned a *bei ya mwisho*, last price, and that is the one the shop assistant needs to know. When a customer asks for the price one usually starts with announcing a sum that lies around 10% above that, hardly more. For a T-Shirt that might be around 1000 TSh, for some material maybe 5000 TSh, with 1000 TSh being the equivalent of approximately 0,53 Euros in July 2009. And, as

Fig. 26: Shops in Macontainer

soon as one mentions the *bei ya mwisho*, the customers either agree or wearily leave the shop. Even when friends come into the shop and beg for a cheaper price by emphasising their close ties to the shopowner, the *bei ya mwisho* is very rarely given up. While this shows that prices are rather fixed and not subject to clever negotiation, it also indicates a rather low profit margin. Whereas in the best case, the traders earn around 10% more, most of the time they get the lowest price needed to make their trading business worthwhile: pay for the flights, the goods, the shipment, the shop and produce a surplus which allows them to look after their families. As numerous examples in Mchangani and Kiponda show, when the volume of sales and/or the *bei ya mwisho* is set too low or handled to loosely, the business easily fails.

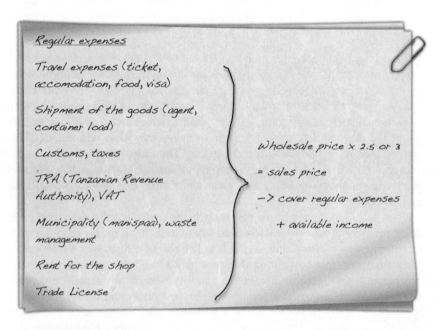

Fig. 27: Traders' calculation

On the other hand, a number of Zanzibari traders who have been rather successful with their trading businesses during the last years, have started to invest their profit by opening a new shop in Dar es Salaam. Whereas in Zanzibar the purchasing power of most customers is too low to allow for a good profit, in Dar es Salaam the sheer number of people already seems to enhance one's business. These investments are generally focused on the area of Kariakoo, the streets surrounding the central market of Dar es Salaam. In the following I will illustrate how the recent development of this area is closely related to the translocal mobility of Swahili traders as well as how this currently very dynamic area offers opportunities but also particular challenges to the traders.

Kariakoo: The place for investing in Dar es Salaam

It was under German rule, that Kariakoo had been re-planned into a grid pattern, which was completed in the 1920s. Still in 1970, Sutton (1970: 13–14) writes that this area, named after the Carrier Corps porters who were encamped there during the First World War, forms 'the central working-class suburb and social hub with the biggest market and main clubs'. At that time, most of the grid consisted of single-storeyed, rectangular 'Swahili houses', and it was only along Uhuru Street and Msimbazi, as well as directly bordering the market, that mainly Indian merchants had some two or three-storey buildings.

When entering Kariakoo today, the picture is completely different. Particularly since 2002, when the markets opened towards Asia (Dubai, and esp. China, Thailand, Singapore), the area has experienced a rapid development, consisting of an enormous building boom and, related to that, the opening of more and more shops. At least 350 high-storey buildings have been built over the last five to six years, providing new business spaces but also thousands of new apartments.

Although generally forbidden by the municipality, some of them, to reduce costs and increase the speed, simply extend and heighten the old houses regardless of their weak fundament. Not being the owner of the land, they are often only given the right of utilisation for twenty years and do not seem to care about the long-term effects of this practice.

Fig. 28: Grid-pattern suburb of Kariakoo in 1969 (Source: Sutton 1970; photo: J. Kirknaes)

Furthermore, as not all follow the rule to let the construction plan be approved by the municipality before starting the building activities, many of the houses are build in such close proximity to each other that, when looking up, it seems as if people are able to shake hands with their neighbour on the tenth floor. With this rapid increase of housing, the infrastructure of the area is completely overstrained. The government decided to improve the roads only after the building activities have calmed down and has applied for the support of the World Bank to also improve the sewage system. Hence, still half of the roads are made of sand, and even those that are not are still far too narrow and, with thousands of cars parking along the sides, it is hard to pass through. Though according to the law, every new-built house has to provide five parking spaces, if as a basement garage, so far nobody has done that. Therefore, at the moment the quickest way to move through Kariakoo often is to walk. Every day, a couple of thousand shops attract masses of people, who all thrust their way through the streets from one shop to the next.

Fig. 29: Kariakoo today (I)

A lot of these investments are made by people from Zanzibar and Pemba, who use the money they have gained from translocal trading activities –together with remittances received from family members abroad – to expand their business to the capital. According to Saleh (2006: 360), in 1999 398 out of 973 shops in Kariakoo were owned by Zanzibari, of which 79% were originating from Pemba. But not only people from Zanzibar and Dar es Salaam, but also from the rest of the country, as well as from Zambia, Malawi and Congo come to Kariakoo to buy goods wholesale. It is especially this big variety of people from different places that contributes decisively to the dynamics and vitality of this place in which diverse translocal ties intersect and moor.

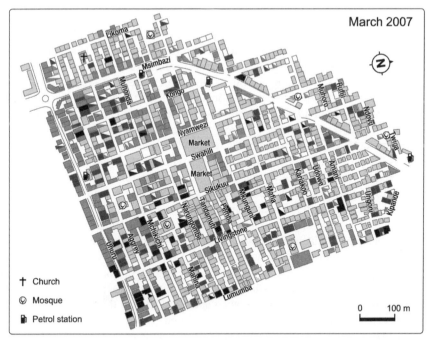
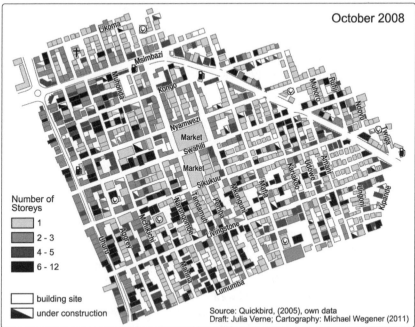

Fig. 30: Recent building activities in Kariakoo

Three years ago, Majid has decided to open another shop in Dar es Salaam and one day, when I arrive in Dar es Salaam having taken the ferry from Zanzibar, Majid picks me up from the harbour to show me his shop. We drive straight on the Morogoro Road through the city centre into Kariakoo. Majid has rented a shop in one of the streets with a high percentage of Swahili traders, many of them originating from Pemba. In Mkunguni Street almost all traders know each other from previous trading activities in Zanzibar or parallel trips to Dubai or Asia. Majid's shop assistant is from Pemba as well and he confirms that this street is like a Pemban enclave. So, why does it make a difference to have this shop in Kariakoo and not in Wete, Chake or Zanzibar Town? While staying in Dar es Salaam and regularly keeping Majid or his shop assistant company in the shop, it turns out that indeed about half of the people coming into the shop are from Pemba. Many of them now live in Dar es Salaam, where they are either employed or engage in their own trading activities. Even if their average income might actually be higher than the one they had in Zanzibar or Pemba, possibly allowing them to be more active consumers, it also seems to be the specific sense of place of Karikaoo that stimulates consumption. Kariakoo is *the* place to go for shopping, it is the place where you find everything; it is the place where growth takes place and where economic success becomes visible. These are the aspects dominating the discourses I overhear and take part in when spending time in this area, clearly serving as decisive motivators to be present in this area.

As Majid emphasises, he was lucky to get a door in this street as it has become increasingly difficult to get a nice shop in a good location at a reasonable price. Whereas he had to pay 300 000 TSh a month for approximately 18sqm in his first year, the rent has now been increased to 500 000 TSh in his third year. Compared to the spaces rented out surrounding the DDC hall, where 600 000 TSh have to be paid monthly for only about 10sqm, it is still a good deal and he hopes that he can stay there for a while. With his shop being located in one of the old single-storey buildings he can never know when somebody might buy the property to invest in a new, multi-storey apartment house. Even if Majid would be offered to open his shop again in the new building, this would probably be at a higher rent and he would definitely have to move out in between.

This already shows, how the rapid development characterising Kariakoo for the last years, despite offering a thriving business environment, also poses particular challenges to the traders. Shops are torn down, new shopping centres are build, old houses are demolished, and new multi-storey buildings are quickly raised, so that the same location might completely change its looks and atmosphere within a short period of time. This leads to a high mobility and fluctuation of shops and people within the area, as traders try to navigate in this dynamic field always with the aim to secure their profit. That this often very unpredictable situation affects the traders' businesses is illustrated well in the case of Saleh.

Saleh, also born in Pemba, has for a couple of years rented a shop at the corner of Swahili Street and Ndovu Street in which he sells car glass. When he hears that in three months time investors will take over the houses on his side of the road and make them leave, he starts renting a shop two streets away to secure a place in the area where he can move to. Although, for the first months he does not sell anything there, he decides to pay a rent of 300 000 TSh on top of his usual expenses to secure himself an alternative location and be able to tell his customers well in advance where they will be able to find him after they will have closed down the building. This way, he hopes to take his regular customers with him to the new location. However, three months later nothing happens, and instead of moving to the new location, he prefers to stay in his old shop situated at one of the bigger roads and thus more easily accessible by customers than the other one. When, six months later, he is told that the investors have cancelled the project he finally gives up the new shop again, hoping that it will take a while until a new investor is found. Although, this second shop has severely reduced his profit for a period of six months, he would probably do it the same way again in case he would be told to move out soon. He would rather pay a double rent again, than taking the risk of not having any shop for a while, and locations that are well suited for trading car glass – while still being affordable for him – have already become rare.

Fig. 31: Kariakoo today (II)

Overall, Kariakoo is probably the most impressive example of a very dynamic area, reflecting much of the material effects of diverse and complex translocal connections in which Swahili traders have come to play an important and strongly visible role. However, as I have tried to show, whereas the area allows for a much higher turnover than any area in Zanzibar, it is not always easy to secure a high profit as the flexible, ever-changing and often highly unpredictable context also brings with it a number of difficulties. These difficulties, such as finding and being able to keep a shop at a reasonable price, can be avoided, for example, when deciding to sell the goods at home. This opportunity is particularly common among female traders who prefer the privacy of the domestic space to crowded shopping areas. By providing an example of a home shop, in the following I will thus give another example of how trading can be organised and is integrated into everyday life of Swahili traders.

Home shops

Although the number of female shop owners has risen a lot over the last years and Macontainer in Zanzibar – where Khadija has her shop – is an illustrative example of women dominated business, a lot of female trading activities still take place out of the public's sight. Maryam is one of these women who have reached their economic independence through selling goods at home. Since 1997, she travels to Dubai every three to five months, depending on how the sale goes. This way, she is able to offer goods throughout the year. Notwithstanding having enough goods to fill a shop, she completely objects to the idea of opening one. Selling the goods at home gives her the freedom to only be a shopkeeper when she wants or when customers are there, allowing her to integrate the business more smoothly into other activities, such as receiving guests or cooking. When she is at home, she is always happy to let customers look at her goods; when she is away, the customers either have to come back at another time or are let in by the housemaid and invited to wait. Avoiding paying a rent for a shop, selling goods at home is also a way to reduce her fixed costs.

I am going to Maryam's house together with two of her nieces who are both looking for a new outfit. Maryam, who has already seen us coming, opens the door for us and guides us upstairs. The female living room in this large house in Vuga, where she lives with her husband and children, serves as the salesroom. The goods are kept in cupboards, on shelves and desks, are spread out on the sofas, as well as on floor. But despite this apparent chaos, Maryam knows exactly where to find things. Apart from female skirts, blouses and trousers, she also sells typical long dresses of Arabic style (*dishdasha*), headscarves (*mitandio*), and the long black overcoats for Muslim women (*mabuibui*). Next to big piles of children's clothes for boys and girls, she also has some shirts and white long dresses (kanzu) for men, as well as underwear. In a glass cabinet, she keeps perfume, frankincense (*udi*) and also incense burner, and half of the big wardrobe is filled with bed sheets, duvets and curtains. She hands us skirts and blouses she thinks we might like, presenting us different colours and sizes. When we want to try something on, we are taken to a neighbouring room to change. In this respect, Maryam's home shop offers a much more private and intimate atmosphere than a usual shop, where there is often only a very provisional fitting room and the whole shopping process is visible to strangers. Sitting on the floor we take over an hour to look through her clothes, being offered fresh juice in between, exchanging news and listening to her business tales.

Fig. 32: Inside a home shop

While we are there, a number of other women, one of them with her daughter, come to join us. It is a familiar, comfortable atmosphere in which we all give each other advise on what suits best. Maryam does well to advertise her products, not getting tired of showing more and more things to us. Over the years, she has got a good impression of what her customers like and therefore tries to meet their taste by sometimes even choosing goods individually. At her last visit, one of the women had asked for a particular kind of *dishdasha*, now coming back to see the newly arrived goods. Being content with Maryam's choice, the woman even buys two of them in different colours.

Although the home shop is generally open to anyone, many of the women coming to Maryam's house have become regular customers. And their number is constantly increasing through personal recommendations. Despite not being visible to the public, Maryam has definitely managed to make a name for herself. Concentrating on higher priced goods from Dubai, she especially draws the attention of Zanzibar's wealthier population (and of those who at least want to appear so). Furthermore, by offering her goods at home in this very familiar atmosphere, Maryam succeeds in creating a particular sense of belonging to an exclusive group.

While Maryam's home shop has become a constant institution and is based on regular business trips to Dubai, many other women sell things at home on a much smaller scale and more irregularly. Either trying to take as many goods when returning from a visit to Arabia or asking other traders to acquire some goods on their behalf, these women take the opportunity to increase their budget without totally subscribing to a traders' life. Often, these women have close connections to other traders whom they can ask to provide them with goods. This also holds true for Massoud's wife Nabila, who tries to benefit from her husband's mobility by instructing him to bring her some goods she can then sell at home. For example, when realising that a new style of headscarves that is just becoming fashionable in Dubai is still rarely available in Zanzibar, Nabila describes them to Massoud so that he can acquire some for her on his next trip. This way, she is able to spread goods of her taste while not even having to be mobile herself.

Overall, looking at the different ways in which traders organise and manage the sale of their goods, one gets an insight into the multiple considerations and calculations that are behind the traders' choices, for example in respect to the employment of a shop assistant or the location of their shop. Whereas some traders like Massoud and Khadija usually spend their whole day in their shops managing the sale on their own, others, like Majid, rather take a kind of supervisory position moving between his different shops and always on the look for new business opportunities. The case of Maryam shows, that domestic spaces also have to be recognised as key sites of trading practices, allowing for a more intimate and exclusive way of trade. In general, it is striking that trading space is often not very separated from other spheres of the traders' life, and that trade is closely connected to other activities, such as spending time on the *baraza* or visiting friends. As I have especially indicated with regard to home shops – but it clearly also holds true for shops in Kiponda, Macontainer or Kariakoo – this trade is strongly influenced by close relationships between traders and consumers. Often, it is customers who explicitly encourage business trips to Dubai and sometimes even place detailed orders and demand specific things. In a next step, I therefore want to turn the attention to the perspective of the customers and their often very active involvement in the translocal trading connections between Dubai and Zanzibar. In doing so, I will contribute to recent (cultural) geographies of consumption by challenging the abstract conception of the shop as a functional, smooth, opaque economic surface and supporting the idea of a more relational understanding of consumption instead that implies to account for the intermingling and entanglements of consumers, shops and goods.

As I have pointed out before, the month of Ramadhan is certainly the most relevant time of the year in respect to the trading business. Generally, it is especially the women who actively drive the trading activities during this time, with most of them having very clear ideas of what they want and being responsible for the selection. To be able to give a detailed insight into the complex negotiations between consumers and traders, I will therefore concentrate in my elaborations on a particular woman. Accompanying her on her shopping tours gives a vivid impressions of how translocal mobility can be demanded – but also bypassed by both consumers and traders – in order to meet the needs of the festivities on time.

DEMANDING OR BYPASSING TRIPS: SHOPPING IN ZANZIBAR DURING THE MONTH OF RAMADHAN

Nabila got married to Massoud in January 2006. Together with their son Naim they live in Michenzani, not far away from her parents' flat where she grew up. Before her marriage, Nabila had been working for a travel agent in Kiponda, but at the moment she feels busy enough looking after their son, doing the housework, and the shopping. During the month of Ramadhan, when Massoud has to spend more hours than usual in his shops overviewing the business, it is in the evening when breaking the fast together that Nabila tells him about her shopping tours through the city, what she still needs to buy in preparation for the festivities and how much money she would like to get from him.

What seams clear is that their son Naim needs at least three new outfits for the three days of the holiday. This usually also applies to adults, with the first day being particularly important as one usually goes to visit relatives and close friends, giving each other 'the hand of Eid' (*mkono wa eid*). While Nabila wants to get a kind of long, wide, Arabic dress (*dishdasha* or *dira*) in a new cut, which she can wear without a black overcoat (*buibui*) on top, Massoud simply needs a new *kanzu*, the long white robe worn by men. This is also what Nabila wants to get for Naim, together with a coat, so that they all look like an Arab family. Nabila and Massoud both belong to Arab 'tribes' (*kabila*) and their Arab orientation is also expressed in the decoration of their flat, the pillows on the floor, the frames on the wall, the curtains...– all being elements that should ideally be replaced by new things for the celebration of Eid-il-fitr.

Our first shopping tour of the day usually starts late in the morning. Apart from buying some ingredients for the iftar, the breaking of the fast, we pass by shops to look for clothes and materials, curtains and carpets, discuss with the tailor the possibility of sewing a coat for Naim in time and visit some of Nabila's friends to exchange opinions about where and what to buy. Being aware of the fact that some things might be sold out fast, already in the first week of Ramadhan Nabila takes me to a shop in Mchangani that is known for its good quality but also expensive materials. While the shop assistants are still busy ordering the newly arrived goods in the shop, we take our time to choose between different colours and designs. Ismail, a former neighbour of Nabila who works in this shop, introduces their offers to us, opening packages and letting us feel the different materials. Without having looked at any other shops, Nabila trusts that the owner of this shop – also a Zanzibari of Omani ancestry – has chosen fashionable material that are worth paying 45 000 TSh to make a good impression on the holiday. According to her, it is important that everyone realises that these are new designs and not the ones that were in fashion last year. So, even though she personally likes a material in burgundy, she can easily resist buying it because it might be considered as last year's style. Finally having chosen our materials, we directly go to the tailor opposite Nabila's flat. He is one of the few tailors whom she assigns the ability to turn the material into a well-fitting dress. Showing him some pictures of the latest designs printed off from the internet and discussing how to combine them with some of his new ideas, he takes our measurements and agrees to have our dresses ready three days prior to Eid-il-fitr.

Where to find the perfect shoes for our outfits is less clear. Almost daily we walk through Macontainer, look at their selection and inquire if any new goods have arrived. While the first things are already beginning to sell out, many of the female traders plan an additional trip to Dubai to refill their shelves by attending to the wishes of the customers. The omnipresent topic guiding these visits to the shops is what is currently in fashion in the United Arab Emirates and Oman, emphasising once again how the translocal links between East Africa and Arabia continue to influence taste and expressions of the self. It is especially Arab material culture in the form of clothing and home furnishing that meets the women's taste and is taken to symbolise their translocal links and distinguish themselves from the 'African' population of Zanzibar. The women's urgent desire to be able to present a newly decorated house to their guests that corresponds to their preferred style can however be used by the traders, who have realised that a pretended translocal mobility of the goods may sometimes be enough to convince the customers to buy the goods.

Fig. 33: Juda's Shopping Mall

After the *iftar*, when Massoud goes to the mosque, Nabila and I often leave the house again to join the masses of people browsing the shops in town. During Ramadhan most shops are open until midnight and strengthened from the evening meal, many use this opportunity to have another wander around in a more agreeable climate. This is also the time when we go to Juda's shopping centre. Juda, a Zanzibari of Yemeni ancestry, is one of the most well-known and most economically successful traders in Zanzibar. About five years ago, he built himself the first shopping centre in Zanzibar opposite the Kwality Supermarket on the road from Michenzani to Mchangani. Whereas his shops downstairs are specialised in female wedding dresses and baby clothes, on the first floor there is a shop selling children's clothes and another one selling items for interior decoration such as bowls, vases, wall frames, mainly of Qur'anic designs, and plastic flowers. In the shop in which children's clothes are sold, numerous mothers with their children try on different shirts and trousers, with the imprint of apparently famous wrestling stars clearly being this year's bestseller. As Nabila is still hesitant to buy something for Naim, we have a look at the vases instead. With Naim having broken the old one when playing football in the living room, she definitely wants to get a new one and Juda's is the place offering the biggest selection.

With the rush of consumers, everything looks messy at the end of the day. A pile of plastic flowers, from which everyone tries to pick the most beautiful ones, lies at the side of one room, making it difficult to pass through to the shelf with the wall frames. Some vases have already been damaged and are now sold at reduced prices. It is clear that soon not many attractive things will be left. However, what we have been told by Nabila's husband, but most of the customers do not know, is that all these goods selling so well are not as new as the customers think they are. Confidentially, Juda had revealed to Massoud that his containers have not yet arrived in Zanzibar, so that he is forced to offer the goods that he still has in his store from last year. By selling the old goods at an even higher price to make them appear more valuable, he successfully plays with his translocal im/mobility, leaving the customers with the impression to acquire new holiday products just imported from Dubai. What might be even more astonishing is that Nabila still decides to buy one of the smaller vases together with some flowers. To her, it is more important that her decoration will be associated with Juda's trading business that is known for importing the latest fashion from Dubai than knowing that, in this case, this is not true. Nevertheless, the name and symbolism of the shops is not always the decisive criteria. On the contrary, Nabila's shopping for new curtains and a carpet is rather characterised by the search for cheap deals.

Being disappointed by the limited choice and high prices of the carpets offered in Zanzibar, Nabila makes an effort to get hold of an old family friend working on one of the boats travelling back and forth between Zanzibar and Dar es Salaam. One evening, we all drive to the harbour before the departure of the boat, so that Nabila can give some money to this man and tell him what kind of carpet she wants him to buy for her. Already the next day, when the boat is back, we drive to the harbour again to pick up the carpet. In this case, what is more important than the (imagined) origin of the carpet is its colour – apparently brown and creamy colours are just becoming very fashionable – which has to fit the new curtains and pillow-cases. Apart from that, even if this did not involve any links to Arabia, being able to get a carpet from Dar es Salaam still expresses that one is well connected translocally, drawing on resources of trading connections and family links elsewhere. While the shiny curtains for the living room have to be bought in a glamorous shop in Kiponda, Nabila, her mom and her friend Sauda all go to

Sateni, a cheaper market to buy new cotton curtains for the bedrooms and the curtain separating the living room from the more private rooms.

As soon as the shopping for the new home furnishing is complete, Nabila's attention returns to her and her son's clothes. By sticking to the approved shops it is relatively easy to find things for Naim. A quick visit to his friend's shop in Kiponda already adds another outfit. What seems more difficult than finding favourable outfits for Naim, is to get a new *buibui* in the particular design that Nabila strives for.

For many years, Arab ways of dressing have begun to strongly inform the way Swahili people dress along the East African coast. Regarding women, the long black veil as it is worn in Oman and the United Arab Emirates as well as in other Arabian countries (called *abaya* in Arabia, but in East Africa usually also named *buibui*) has widely replaced the traditional Swahili *buibui*, and has become the dominant overcoat of women in public (cf. Ivanov 2010). This long black dress can be made of different materials, adorned with different designs in different colours and sawn in different cuts: wide, fitted, with wide, slim, or trumpet arms, like a butterfly, 'Islamic' style or with a v-neck. At least every year a new style is introduced, generally created in the fashion stores of Dubai and other Arabic cities. And with only a short time-lag the new style can also be seen in Zanzibar. Not simply buying one of the old designs but actively following the latest Arab fashion and participating in setting the new trend in Zanzibar is a clear way of showing ones Arab affiliation, origin and identity. Nevertheless, with the new designs being rather expensive, ranging from more or less a hundred to several hundred US-Dollars, it is not possible for many Swahili women to import them from the Arabian Peninsula. Moreover, as these kinds of *buibui* are usually made individually according to ones measurements, it would need the help of relatives or friends abroad to get one made for them.

With these thoughts in mind, Nabila therefore decides to get her new *buibui* made in Zanzibar. In Vikokotoni we find some material she likes and in some shops specialised in *buibui*-accessories, we also get shiny stones for small ornaments. When we have everything together, we go to her favourite tailor. Although Nabila has a clear image in mind of how the *buibui* is supposed to look like, it is not easy to explain that to the tailor. All the veils he has finished recently do not match her ideas and after a while they agree, that Nabila has to find a model of her desired design, which he can then try to copy. This now means to activate all her contacts to get hold of such a *buibui*. For two full days Nabila is busy calling her friends describing the design she has in mind and asking whom they know who has such a *buibui*. Unfortunately, just one of her former colleagues already own this style, but Nabila has never been very close to her, so that she would need another friend to ask for borrowing it on her behalf. At the same time, she motivates some other friends to look for this style in their environments. But, the first *buibui* we go to see is not the style Nabila is looking for, and, the second one we find out about is owned by a woman who does not want to give it to us (it seems the women prefer to be unique). Finally, three days later, Nabila manages to borrow a *buibui* from a friend of a friend, which is sewn almost exactly the way she wants hers to be and takes it to the tailor. After telling him what to do differently and how to put the ornaments, she returns home full of enthusiasm. But only one week later, when we go to pick up her *buibui*, the discussions about the 'right' design start again. At first, Nabila is very critical with the result, questioning the tailor's work with regard to the way the *buibui* falls at the back. Finally, some of her friends can convince her that it looks like the design she wanted and will surely be considered as the new Arab fashion and that, above all, it definitely fits her very well.

Fig. 34: Dressing up for Eid-il-fitr

As this case shows, it is rather the imagined than the actual origin of the good that counts to the consumer. Apart from buying clothes imported from Dubai, the creation of copies still symbolising Arab fashion is another way of living and performing these translocal connections. This way, dress as well as interior decoration is used to demarcate their belonging to what is considered as Arab culture. Through taking part in 'Arab' fashion these women fashion Arabia. The result is a highly fashion conscious environment, full of what in Kiswahili is popularly termed Bi Show ('Ms Show'), always eager to express their translocal identity through having the latest styles and designs.

CONSUMPTION FROM A SWAHILI PERSPECTIVE: REFLECTIONS ON CULTURE AND ECONOMY (II)

Geographies of consumption, similar to the earlier mentioned retail geographies, have long been considered as a rather boring field of geographical enquiry. Having often been subject to a one-sided view of consumption as either good or bad, it is especially over the last ten years, that a more considered analysis and reflection has taken place. As Crang and Jackson stated already in 2001, 'old oppositions between consumption and production, between pessimistic readings of the consumer-as-dupe and optimistic readings of heroic consumer resistance, have been transcended' (Crang & Jackson 2001: 328). Apart from studies on 'lifestyle shopping' and 'heroic consumption' in mega-malls (cf. Shields 1992, Ley & Olds 1988), what has become popular – particularly in British work – is a focus on the relationship between consumption and identity in more mundane or alternative settings of everyday consumption (see Jackson & Moores 1995 on domestic consumption, or Gregson & Crewe 1994 and Crewe & Gregson 1998 on forms of alternative consumption, and Miller 1998 on a theory of shopping more generally). Nevertheless, the various ways in which researchers attempt to empirically rethink the relation between consumption and production – as well as more generally between culture and the economy – are often still dominated by more abstract methodological and ideological discussions. As Jackson pointed out, 'it could be fairly said that calls to transcend the "great divide" between the cultural and the economic have significantly outnumbered empirically grounded studies that demonstrate the difference that such a move would make in practice' (Jackson 2002: 4). Although this claim has already been made eight years ago, and a number of intriguing empirical work has been done since then, recent publications still show clearly that the debate between those reducing the economy to either political or profit-maximising logics (i.e. 'political economists' and 'economists') and those understanding economic practices as always being embedded in and contributing to cultural contexts is still going on, often devolving into, as Heuman (2003: 104) has called it, 'a theatre of polarised argument and mutual misrepresentation' (for illustrative examples see e.g. Castree 2004, Gibson & Kong 2005, Heuman 2003, Jessop & Oosterlynck 2008, as well as the progress reports on geographies of consumption by Crewe 2003, Goss 2004, Mannsvelt 2008, 2009, 2010 in *Progress in Human Geography*).

Without wanting to deny the relevance of some of these discussions centring on the advantages and disadvantages of an increased attention to the mundane and particular, or the special emphasis on the entanglement and inseperatability of culture and the economy, I still do not think that this debate, due to the one-sidedness of the respective positions, has been very fruitful. By arguing mainly on ideological grounds, the debate often lacks a deeper engagement with the concrete contents of empirical research.

Following the flows of people, goods and ideas between Zanzibar and Dubai, it was my aim to better understand the organisation of these links that has drawn me into geographies of consumption. First, this section has therefore illustrated how shopping forms an important part of the trading connections. Instead of the consumer being the receiver at the end of a one-directional connection, certain practices of shopping are not only enabled by the translocal trading connections, but are also explicitly demanded by posting orders. While the translocal connections of the traders are a necessary precondition for offering goods – and hence for their economic success – translocality, thus, is also crucial on the consumption side, as the actual and/or imagined origin of the goods plays a decisive role in satisfying the consumers' needs and demands. Moreover, ways of bypassing and playing with this translocal mobility, such as, in Juda's case, inconspicuously offering his old products, or Nabila getting made a Dubai-style *buibui* in Zanzibar, show how translocal connections can be evoked without actually having taken place. Secondly, the previous pages have exemplified how an examination of shopping practices points at the meaning of these translocal connections for processes of identification, expressions of belonging and attachment. In Zanzibar, shops as well as particular goods act as symbols of the translocal connections between 'Arabia' and 'Africa', expressing and constantly recreating one's translocal belonging. As Leiss et al. (1990: 304) have argued, 'consumers [...] do not – nor do they wish to- form their tastes and preferences in the private bliss of ratiocination and then descend upon innocent merchants to scrutinise their shelves with cold and wary eyes', but instead they are often very closely connected to the traders and constantly exchange ideas, concerns and demands. In this respect, not only the practices but also the discourses of consumption preferences are generally laid open to both traders and consumers.

In an often quoted example of geographies of consumption in which May researches the discourses of consumption among the inhabitants of Stoke Newington in London (cf. May 1996a, 1996b), it is pointed out that consumption and particularly food consumption might serve as a way to 'eat the other' through striving for ethnic variety and the exotic (cf. hooks 1992). On the contrary, as this research shows, the goods bought in the shops in Zanzibar and Dar es Salaam are actively constructed as 'self' although imported from far away. With the goods being 'as Arab as oneself', consumption, here, can rather be regarded as a way to reinforce and strengthen one's cultural identity and belonging through buying 'the self' and not 'the other'. This also clearly illustrates the territorial ordering and bordering functions of everyday consumption practices, which in this case however do not demarcate a closed physical entity but help to establish a translocal space. By doing so, the ethnography of shopping adds a crucial perspective to the meaning of and engagement with translocal connections in everyday Swahili lives. Furthermore, going beyond the activities of the traders, the extensive shopping activities during the month of Ramadhan serve as a valuable lens through which to understand the dynamics and interplays between trade and consumption practices, illustrating how translocal connections are lived, demanded, and also played with in the construction of ideal careers and identities.

4TH HALT: LIVING (UP TO) THE TRANSLOCAL IMAGINATION

As this chapter has shown, traders like Majid, Khadija, Maryam, Massoud and many others today try to make their living by engaging in international trading practices, importing goods and selling them in Zanzibar. Frequently travelling to Dubai, they are all constantly on or in-between trips, planning or preparing for the next one, while at the same time still occupied with the previous one – especially by way of anxiously waiting for the arrival of their luggage. Although still using the same route as the early Indian Ocean traders and seafarers who have connected 'Arabia' to the East African coast, they have now become business travellers, reducing their own travel time to less than six hours, while their luggage crosses the Indian Ocean in one of the container vessels which have slowly but steadily come to replace the old dhow trade. While, around sixty years ago, these trade journeys have lasted for months, many of the traders now only travel for a couple of days turning these long journeys into short and efficient business trips.

Looking at Dubai from a relational perspective, its establishment and importance as a 'shopping paradise' is strongly related to its position as a node of global financial and material flows in which the connections to East Africa only play a minor role. Nevertheless, Swahili as well as other East African traders are also part of these flows, and it is through their regular mobility between the East African coast and Dubai that this connection develops important material effects in both places. In respect to the Zanzibari traders presented here, it becomes clear how their high mobility also demands certain material fixations, such as the houses of relatives in Dubai or a place like the Gold Plaza Hotel from where to organise their shopping tours. Besides, the imported goods require a place to be stored and sold, so that all business trips are also linked to phases of shop keeping. As already pointed out by Constantin and Le Guennec Coppens (1988) in their article *Dubai Street, Zanzibar...*, being the source of a high percentage of the goods, 'Dubai' has become materially visible in the shopping streets of Zanzibar.

Furthermore, it is particularly these connections between Zanzibar and Dubai that have recently triggered the rapid building development in Kariakoo, Dar es Salaam. For many Zanzibari traders Dar es Salaam has become the place where they hope to increase the sales rates, the place for new investments, while Zanzibar – at least with respect to larger businesses – seems to turn into a side scene of their trading activities. However, the very dynamic and often unpredictable development of Kariakoo also imposes particular challenges to the traders, sometimes even leading to severe reductions of their profits. The shops, if in public or domestic space, can be regarded as the material effects of translocal trading activities, with the goods on display being like the footprints of this mobility that reveal a lot about the direction of the connections and the status and ambitions of the trader. How closely the trader's financial success and the expression of their translocal connections to Dubai are intertwined can be seen when turning the attention towards the consumption practices in Zanzibar.

Particularly during the month of Ramadhan, the economically most relevant period of the year, the actual or metaphorical translocality of the goods is the crucial basis for the traders' economic success and pays off culturally as well as socially. Imagined and actual family relations are connected to imagined and actual business spaces, with one implying the other, mixing, overlapping, and enforcing each other, sometimes even if not actually existent. From the ways in which the goods are arranged, over their accentuated quality and origin, to the special bags in which the goods are handed over to the customer – all expresses the claim to belong to a community of translocal traders that live up to the translocal imagination of many Swahili.

Having become a business traveller to Dubai is indeed widely regarded as pursuing the ideal way of live in the Swahili context in which discourses about historical and contemporary mobility are vital to an understanding of self. While, on the one hand, it is the existing translocal links that – depending on the involvement of family members – either actively facilitate or at least discursively frame their business endeavours, on the other hand, it is also their mobility and the links they create that influence and strengthen the translocal connections. Accompanying this mobility and trying to understand their everyday lives on and in-between these business trips illustrates, how these traders not only actively live but also enliven – materially and imaginatively – the translocal imagination that is so important for processes of identification and senses of belonging today. This ideological dimension of trading practices will be at the centre of the following chapter, in which I will now turn the attention to the everyday lives of less economically successful traders.

STICKING (TO) TRADING CONNECTIONS: OBJECT GEOGRAPHIES THROUGH THE HANDS OF WANNABE TRADERS

PRELUDE: FOR FOLLOWING OBJECTS

When looking at the trading activities of the majority of Swahili traders, it is much harder to distinguish clear routes and connections, which can be accompanied by the researcher, than it has been in the case of the more professional traders that have been at the core of the previous chapter. The majority of Swahili traders work on a much smaller scale, over shorter distances and could rather be called occasional than professional traders. Nevertheless, even though these traders hardly travel any great distances themselves, they still connect people and places and thus form an active part in the translocal Swahili connections.

To find out about the organisation and meaning of this kind of trade, I started by accompanying some of the young people who told me that they are currently engaging in the trading business. Soon, I was led into a rather messy setting, full of different goods, unclear and confusing deals with a variety of places involved, with people always on the move but not always sure (at least from my perspective) for what reason or if with any financial success at all.

> 'Goods flow through the market channels at a dizzying rate, not as broad torrents but as hundreds of little trickles, funnelled through an enormous number of transactions.' (Geertz 1963: 31)

Although it was exactly this messiness that interested me, I saw a danger in getting lost in these multidirectional and complex flows and therefore decided to focus on particular things and their paths, which could serve as an orientation from which to explore this kind of trading practices and their everyday context. By concentrating on specific commodities, I was then able to shed light on the practices involved in their trade in order to gain a better understanding of the ways in which their trade is organised, how they move from one person to the next and how this relates to the specific social and cultural contexts of the traders.

This resonates with a recent interest in 'follow-the-thing' geographies in which it is particularly the biographies and geographical lives of commodities that have gained geographers' attention, as they are seen to provide a unique window through which to understand their social and cultural contexts (see also *Doing mobile ethnographic research*, p. 56). By bringing together human and object mobilities through following the movements of particular things, I gained deep insights into the diverse encounters between things and people on the move in a context in which these mobilities are much more individual and variable than in the more professional and larger business endeavours presented before. In contrast to the work that aims at 'rethinking the object' by putting a special emphasis on the meanings of objects as 'not only plural and contested, but also mutable over time and space' (Bridge & Smith 2003: 259), in this case, my approach consisted particularly of following the actual movement of objects. Seeing objects as active constituents of social and cultural relations, this allowed me to explore the connections through which mobile objects move and, by doing so, to gain access to the translocal lifeworlds of occasional traders. Although not accompanying traders on a journey or trip on one distinguishable connection, this chapter therefore follows the flows of a number of objects from Southampton in the United Kingdom, to Dar es Salaam, through Zanzibar and Pemba. This way, I will be able to exemplify the ways in which goods are acquired by traders, how both sellers and buyers are related to the traders and what kind of mobilities are involved in the deals. These, as well as numerous processes of reselling and their discursive contexts, hint at the complex intertwinings of culture and economy, of 'cultural baggage' and 'luggage'.

Whereas economically, these trading practices often do not seem to pay off at all, it is their cultural significance that has to be considered as a crucial driving force of the activities. In this respect, being attentive to the ways in which the meaning of trade is discursively constructed shows how trade in the Swahili context can be understood as a cultural ideology that translates into practice even in cases in which there is a clear lack of economic success. Overall, this chapter thus opens the view to the majority but still less visible group of Swahili traders and their ways of living (in) and experiencing the translocal Swahili space.

MIGRATION AND DEVELOPMENT?
RECEIVING GOODS FROM THE UNITED KINGDOM

About every three or four months Badi and Manju receive a container full of goods from one of their cousins living in Southampton in the United Kingdom (UK). They do not know exactly in which city their cousin lives or what else he does for a living, but they have his number and eagerly wait for his call to hear when the next container has been sent on its way to Dar es Salaam. At least this is the way things have worked for the last half of the year. Six months ago, their cousin had finally agreed to do something to support them in their attempts to establish their lives as traders; and instead of sending money he now sends containers full of used electronic goods for them to sell in Dar es Salaam. While he tells them the amount he wants from the sale, Badi and Manju share the sum they gain on top of that. This way, according to the two young traders, it is a good deal for all of them. They trust their cousin that he more or less demands the amount of money he needs, to have enough to pay for the next container of goods and its shipment, but they have no idea how much this is exactly. In their estimations they count with a sum of approximately 2 500 GBP that they are supposed to send to the UK. Of course, they will only be able to return this sum after the successful sale of the goods. But before that, there are other things to do and pay in order to, first of all, get hold of the goods.

As soon as Badi and Manju, who were both born in Pemba and have only recently moved to Dar es Salaam to benefit from the busy trading environment, get the information that a container for them is heading towards Dar es Salaam, they try to activate their connections to the harbour. Although the container's journey from the UK to East Africa generally only takes about three weeks, one time, it has taken additional two months until they were allowed to pick up their goods, as the clearance as well as the taxation took so long. The first time, Badi and Manju had to borrow some money from some friends and an uncle in order to be able to get the goods out of the harbour. The next time, they were already able to pay this from the money they had gained from the sale of the received goods. And this time as well, they hope to have made enough profit before the next container arrives. Knowing more people at the harbour, and with a better knowledge of the general procedures, they also hope to be able to reduce some of the costs and the waiting time. Moreover, when awaiting the first container, they also had to borrow some money to pay the first rent of a space where they could store and display the goods. This is another expense they are now able to cover from their sales. To be in a central location to start their trading activities, they chose an empty store in the back of some shops in Kariakoo at the corner of Swahili Street and Nyati Street (for an overview of Kariakoo see map on p. 127).

Although the last container has already arrived eight weeks ago, the warehouse is still packed with goods, the high majority of them being used electronic devices. Between a few big refrigerators, there are high piles of DVD players and video players, computers, televisions, rice cookers, electric irons and a few stereos. One of them is connected to electricity, so that Badi and Manju can listen to music when they are waiting for clients, (re)arrange or even repair some of the goods in the store. Although some of the goods look almost as new, others have a much older and used appearance with little scratches or damages. Some even do not work at all at the moment but need some repairs. Whereas sometimes it is worth repairing them and sell them afterwards, in respect to most of the broken goods they just wait and see if they find a customer who buys it at a very reduced price and repairs it him/herself. Despite it being a little dark in the store – Badi and Manju still have not found the time to install another bulb – the place has a busy and active atmosphere. Some friends come and go, plans are made of how to improve the store, and discussions are going on about which kinds of goods are most easy to sell these days. At the moment, it is especially the goods that are less shiny that still fill the space of the store, but Badi and Manju are aware that these either have to be sold or (in the worst case) given away before the next load of goods will be received in order to have space for the new arrival.

This example of trading connections between the two young men in Dar es Salaam and their cousin in the UK clearly fits the idea of a strong relation between migration and development, which has become more and more prominent in development discourses as well as in academia over the last years (Castles 2009). Numerous initiatives have been founded that try to foster a link between migration and development as this is regarded to 'have the potential to generate a win-win situation for all parties' (Blion 2002: 235). Activities of migrants to invest in or support their relatives, friends or organisations in their country of origin are generally very positively perceived by development institutions as well as by many researchers, as migrants are expected to know what is most needed. Due to their close connections to the people, they are supposed to more easily fulfil the ideals of participation and sustainability. Nevertheless, it has often been stated that more information is needed about the motives, objectives, contents, and impact of these activities in order to be able to assess their contribution to and role in overall development aid (cf. Blion 2002). In this respect, Goldring (1999: 163–164), for example, holds the opinion, that the primary reason for sending money or investing in other ways in their places of origin is to increase their own social status, especially in situations when they are not well received and valued in their current host community. Furthermore, Goldring points out the aim to also increase the social status of the community more broadly, corresponding to the idea that migrant communities often have a marginal status in their home countries (cf. Goldring 1999, de Bruijn et al. 2009). It is said that through their investments they want to influence and modify the social, economic and political geographies of their place of origin in order to create a higher consistency between this place and their anticipated and wished-for status. Besides, others have pointed out the clear search for profit as the central criterion for selecting activities (cf. Castles 2009). Generally, both positions emanate from the idea that there must be clear reasons, personal calculations and rational choices motivating these activities, suggesting knowledgeable and well thought-through projects. The presented case of Badi, Manju and their cousin in the UK nevertheless draws a rather different picture.

According to Badi and Manju, it was them who were finally able to convince their cousin that he could essentially contribute to their well-being by providing them with goods. Not knowing the specific situation of their cousin, they simply assumed that there would be an opportunity for him to acquire low priced goods, which are demanded in Dar es Salaam and send them over. While other relatives in the UK were more reluctant, this one eventually agreed and soon after managed to send the first container. Even when, two months after the arrival of this first container, Badi and Manju had not yet paid the whole 2 500 GBP they had previously agreed upon, the two of them convinced their cousin to continue, arguing that their financial situation would improve as soon as they would have returned their other debts. At that point, the cousin brought up the idea to import original electronics from China instead of buying used ones in the UK, but in the end he postponed these thoughts and sent another container of goods from England. Due to less capital, the overall quality of the second load of goods was considerably lower, but with more experience and more connections on the side of Badi and Manju the sale went relatively well and, knowing that this could otherwise mean the end of their young trading activities, they sent back more money than the first time.

As it is clear that the active encouragement to send containers full of used electronic goods to Dar es Salaam came from Badi and Manju's side, it is harder to say what actually motivates their cousin to agree to this. With respect to the social status, referred to so prominently by Goldring (1999), it can be stated that, compared to building a house, sending a car or paying for a funeral or wedding, this kind of investment is a lot less visible to the public. In contrast to buying a (longer lasting) thing or paying for a (widely perceived) event, he invests in a practice, i.e. trading. As Badi and Manju claim, if it would not bring him any benefit, he surely would not do it, but they have no idea what the concrete benefit might be. In their opinion, it cannot be a big effort for their cousin to get the goods, as they expect him to know where to acquire cheap used goods in his place, as they do know that in Dar es Salaam. But that the whole endeavour is not solely to make a profit on his side becomes clear when hearing that, together with the start of these trading activities, he opened a UK bank account for the two of them where they are supposed to save their surplus while only keeping enough money to cover their daily expenses. This expresses the wish to help his cousins in East Africa on a sustained basis to establish their lives, being in line with dominant development discourses. Moreover, although after the first two containers, the cousin has only received 3 000 instead of 5 000 GBP, he still continues to send a third load.

Unfortunately, even though I tried, I was not able to meet this cousin during my stays in the UK. However, having led many conversations with Swahili on this exact topic, I would be surprised if the motivations of this cousin to establish material connections to his relatives would not be as ambiguous as in the other cases, consisting of a complex interplay of emotional, material, social and cultural dimensions that are not easy to resist and often lead to activities far less goal-oriented and thought-through than usually assumed in the literature. In order to better understand the motivations of these investments in trading activities and their role in respect to the social status of the people involved, it is necessary to take a closer look at their impacts on and the ways in which they are negotiated in the Swahili context.

THE EVERYDAY LIFE OF *MASELA* – FOLLOWING 'URBAN SAILORS' AND THEIR GOODS

Taking a closer look at the goods and their ways from the harbour to the warehouse, until they finally reach a new owner, it is striking what a multitude of people is involved in their trade. Apart from Manju and Badi who often rather take a supervisory and managing position, a number of their friends are more actively involved in the selling of the goods. Two of them are Matar and Seif, both oft hem also coming from Pemba. They all went to school together and have spent most of their time in the same neighbourhood in Wete, in the north of the island.

After failing the exam that would have allowed him to continue with Form III – which would have been the 10th year at school – Matar has spent a couple of years in Zanzibar. But not having had any regular job he moved back to Pemba where he started to engage in some small trading activities with his friends. Sometimes helping out others to look after their shops, taking a shift of a related minibus-driver, or delivering goods to the countryside he managed to keep himself busy, but as soon as an opportunity arose he kept travelling to Zanzibar and sometimes also to Dar es Salaam to hang out with some friends and look for business opportunities over there. It was on one of these trips that Matar met up with Badi and heard of their container trade. Accompanying him and Manju on their trips to the harbour and helping them to get the goods to the warehouse, he got more and more involved and simply stayed on. Meanwhile, he has moved in with another Pemban friend, working as a pilot in the harbour of Dar es Salaam, and his wife, who rent a flat in Ilala.

Seif, on the other hand, with both parents having died early, was sent to live with some relatives, first in Tanga and later in Morogoro, during his teens. Recently, his uncle decided that it would be best for him to move to Dar es Salaam where he is supposed to help him in his business. His uncle owns a small logistics company and wants to have Seif as a driver to distribute things within the city of Dar es Salaam. So far, Seif has nevertheless only been called very irregularly as his uncle first needs to acquire a new vehicle for him. In the meantime, he is left to engage in some other activities and was therefore keen to get involved in Badi and Manju's container trade. As he does not want to stay with the family of his uncle, Seif shares a room with his younger brother Khamis in a 'ghetto' in Kariakoo.

The term 'ghetto', in this context, is generally used to refer to a place occupied by unmarried young men (cf. Raab 2006, Remes 1998). Surrounding the courtyard of a typical Swahili house in Muhoro Street, Seif, Khamis, and six others have rented four rooms, while the owners – a Pemban family – live in the main building. Abdulrazak, the only one, who is from Zanzibar and not from Pemba, only spends a couple of days each week in the 'ghetto', as he is working for one of the ferry companies and therefore permanently moving between Dar es Salaam and Zanzibar, where he has his main residence. Hilal and Said, who share the biggest room, have come to Dar es Salaam to study information technology. As soon as they have holidays, especially Said travels back to Zanzibar and Pemba to look after his shops, one selling mobile phone accessories in Zanzibar and another one selling spare parts for motorbikes and scooters in Pemba. For this reason, even when studying, he also uses the opportunity of being in Dar es Salaam to look for cheap deals for his own trading practices. Apart from these three, the other five are typical 'urban sailors' (*masela*), all involved in some rather irregular trading activities, always searching for a good deal, helping out here and there and trying to get a share of the benefits.

The expression *msela* (pl. *masela*) is commonly used among young, mostly unmarried men, with a positive connotation indicating coolness and a street-smart attitude in the mastering of the challenges they have to face, esp. in respect to earning a living (cf. Remes 1998). Here, *msela* is usually used as a name for friends. When used by others, however, *msela* often means a lack of social acceptance and refers to the wild life of a bachelor often associated with vagabond and immoral behaviour (*uhuni*) (cf. Beck 1992: 129). Whereas Reuster-Jahn and

Kiessling (2006: 49) refer the term *msela* to the older Swahili word *sela* (hooligan), others see its etymology in the English term sailor (cf. Raab 2006). Considering both how the term is used by the young men themselves, mainly in the urban contexts of Zanzibar Town and Dar es Salaam – where it also featured prominently in the lyrics of Swahili HipHop songs –, as well as the lifestyle it refers to, the translation into 'urban sailor' seems to me to best capture its meaning. Nevertheless, the way the term *msela* is discussed by the wider society – and for the term 'ghetto' we find very similar connotations – also indicates the ambivalent and ambiguous nature of their social position. Regularly confronted, even if only implicitly, with the doubt of the older generations that they will ever make it, and criticised by girls for not yet being able to afford a wife, they all feel the pressure of proving themselves.

Despite living together, sharing rooms and helping each other out in cases of emergency, each of the eight young men is first and foremost concentrating on his own activities. Although this might sometimes also lead to a joint business deal, they generally all mind their own independent busyness, a concept that is well put in the Swahili expression *kila mtu ana mishe mishe zake* – everyone is concerned with his own missions. The emphasis on the plurality of missions already indicates the patchworking that characterises this kind of trading activities and will become clear when looking at the everyday lives of Matar and Seif as they are trying to sell the goods that are left in Badi and Manju's store.

Kufuatilia vitu, kuwafuata watu... – Attending to things, following people...

Generally, Seif gets up around 9 o'clock in the morning and, depending on his or his brother's financial situation, he sometimes buys some bread or *vitumbua* (sweet rice cakes) for breakfast but more often leaves the house with an empty stomach. From the 'ghetto' in Muhoro Street, it is not far to the shops of his friends Anwar and Sleyyum where Seif usually passes by in the morning. Sitting together with them in the entrance of a shop, they observe their surrounding, comment on what is going on, greet people they know, and exchange the latest news. Seif is especially alert if somebody mentions someone who is interested in buying something that he might be able to provide from the store or from some other friends who are selling goods. After a while, when the talkative and active atmosphere at the start of each day has faded to a certain monotonousness of mundane shop keeping, he walks up to the warehouse to see Badi and Manju who are usually already there, arranging and rearranging the goods and discussing their business plans. Sometimes, they all spend up to a couple of hours just sitting there, listening to music, repairing some of the goods, negotiating prices and waiting for some customers who have signalled that they would pass by to have a look at the goods. Nevertheless, in contrast to the usual shop keeping, this kind of container trade generally affords a higher mobility of both goods and traders. As there is no shop which is generally visible and open to everybody – be they acquaintances or strangers –, the central task of the traders is to either bring the goods to interested customers or to bring potential customers to the warehouse.

Usually, both Seif and Matar use the time of the late morning to go and see people they know or have heard of who expressed an interest in buying a particular good. These days Matar tries to sell one of rice cookers still left in the store from the previous arrival. It is one of the items that are still in a good condition, so that he hopes to easily find a buyer. Besides, especially as Badi and Manju urgently need more space in the warehouse to accommodate the next container load, they are willing to sell things at a rather low price, so that Matar aspires a good profit for himself. The only problem is, that buying a rice cooker often follows the wish of the woman doing the cooking, who generally is a lot less accessible for Matar than the male members of a household. He has therefore started at his friends' places where he knows their wife and can more comfortably ask them if they want to buy a rice cooker or, rather, if they want their husband to buy them one. Moreover, he has talked to some other friends encouraging them to think about such a purchase, playfully mentioning how this would help their wife and would thus also be to their benefit. Realising that he needs to be careful not to cause any serious discussions among husbands or among husband and wife about the tasks and workload of a 'good wife' or the duties of a 'good husband', which might rather lead to a fight than to the selling of the rice cooker, Matar today decides to concentrate on families who already have a rice cooker but might want a better one.

Fig. 35: Masela *in front of the warehouse*

Already here, it becomes clear how specific goods require specific strategies and ways of trading them. The strategies applied by the traders to navigate in these specific settings often derive from their own experiences and ideas of possible reasons and contexts for buying things and are therefore mainly applied to people they know or who are at least somehow connected to them. It rarely happens that things from Badi and Manju's warehouse are sold to anyone who has no connection at all to any of them. Whereas on the one hand, this is related to the trust customers need to have in the traders not to be sold anything that is covertly broken or of very bad quality, on the other hand, this is also important with regard to the traders' confidence in knowing how things can be advertised and sold.

For these reasons, Matar goes to visit an old friend at his shop in Ilala today. From the last time when he was invited for dinner he remembers that he and his wife have an old rice cooker at home, and that his wife keeps complaining about it. When Matar brings up the issue, he gets told that the rice cooker is indeed not working anymore. But with the help of a new housemaid his wife has started to cook the rice with charcoal again which produces their preferred taste, so that – at least for the moment – he is not interested in buying a new one. Neither has he heard of anyone else who is. Being a little discouraged and hungry, Matar takes a minibus back to Kariakoo, heading straight to Livingstone Street where he hopes to be able to borrow the motorbike of a friend who works in one of the Bureaus de Change in that area. On his way, in Mkunguni Street, he passes by some friends, among them Seif, who stand by a small shop specialised in selling Arabic perfume. The shopkeeper is a friend of Seif whom he also knows from Pemba. Recently married, he has invited Seif to come home with him for lunch to be introduced to his wife and new home. So far, Seif has not been successful in arranging a deal either, still trying to find a buyer for one of the stereos still left in the warehouse.

Fortunately, Matar can take the motorbike of his friend to drive to his older sister who lives in Buguruni for lunch. Accepting her usual teasing concerning his mobile lifestyle, always arriving as suddenly as he leaves, and depending on her to get something to eat when no other opportunity has arisen, he is happy to relax for a while, drowsily watching TV after having eaten. Complaining about the difficulty of selling a rice cooker while at the same time waving the topic aside not willing to engage in any discussions on his current state of affairs, it is his sister who improves his mood by telling him that one of her friends has only yesterday mentioned an interest in buying a rice cooker. Since this friend will come to visit her again tomorrow, they agree that Matar will come back later to bring her the rice cooker he is trying to sell so that she can show it to her. Hence, they part in a far more positive mood, his sister showing to be a little proud of being able to help her friend, and Matar himself now seeing a successful transaction finally coming nearer. On his way back to Kariakoo, he then bumps into a friend who, after the exchange of the usual greetings, directly asks him if he knows anyone who has a good stereo to sell. They agree to meet later at Badi and Manju's warehouse where he can have a look at the ones in there.

After having spoken to Badi on the phone to tell him that at least one of them should be there later to open the warehouse, Matar also informs Seif and then takes the opportunity of having the motorbike at hand to quickly drive to another friend in Kinondoni where he had left the rice cooker two days ago when showing it to a friend nearby. From there he directly drives to his sister's place in Buguruni again, where he uses the chance to talk to his brother-in-law, telling him of the things they still have to sell so that he can spread this information among his friends. Only when his friend calls him to ask for the return of the motorbike, he leaves the verandah in front of their house and drives back to Kariakoo. From Karikaoo he gets a lift from one of his friends also living in Ilala, so that he can get to his place and have a shower. But he does not stay for long, calling Badi to pick him up with his car so that they can get back to the warehouse.

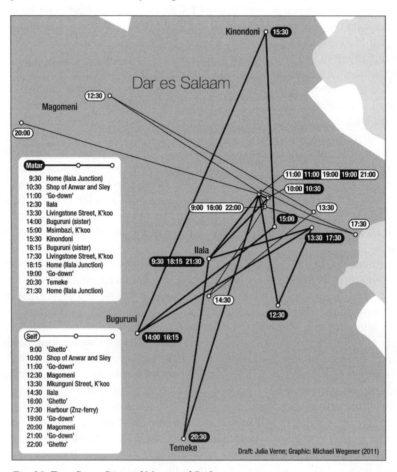

Fig. 36: *Time-Space-Diary of Matar and Seif*

At about 9pm, the friend who is interested in buying a stereo arrives at the warehouse accompanied by another man. The two of them have a look at the different models still available, discussing their qualities and facilities with Seif. They soon have a favourite one and start negotiating the price. However, in the presence of

Badi and Manju it is not easy for Seif to bargain because he knows that for them it is more important to sell the stereo at all than for Seif to make a good profit by selling it at a higher price than the amount Badi and Manju have set as the minimum. And with Matar having brought the customer, he will be entitled to a share as well. In the end, the customer agrees to a deal in which Matar and Seif only gain 5 000 TSh (approximately 2,70 €) each.

After having eaten some 'chicken and chips' together, the other three take Badi's car to drive around and pass by some friends in Magomeni. Matar instead takes Manju's Vespa to drive to Temeke as his host Hamud has told him to pick up some money from a friend which Hamud needs to give to Matar's cousin who regularly travels to Dubai to trade in cars. This time, Hamud wants him to buy a car for him as well that he can sell in Dar es Salaam, but so far, he has not received the amount promised to him by this friend in Temeke. Although the last time Matar has been sent to ask him for the money, the person had told him that he will definitely give him the money this week, today again he is put off to the next week, when he is expecting to receive some income from somewhere else. Driving back to his place in Ilala, he receives a call from Seif who tells him that they have sold another stereo. This means that, apart from having failed to gain something from driving to Temeke, he is now also missing out on the profit of this spontaneous deal. Before going to sleep, he discusses with Hamud where else he could get the money necessary to take part in the car trade. They decide that Matar should go to ask another friend who has a shop in Kariakoo the next morning when on his way to Badi and Manju's warehouse.

When comparing the everyday life of these young traders to the observations made by Geertz during his research on 'the bazaar type economy' in Modjokuto, Java, between 1952–1954, it is striking that one can in part indeed draw very similar conclusions. Regarding this kind of container trade in Dar es Salaam, one can also state that it is based on an 'incredible number of ad hoc acts of exchange' (Geertz 1963: 29) and that the 'flow of commerce is fragmented into a very great number of unrelated person-to-person transactions' (Geertz 1963: 28) in contrast to being firm-centred. It often seems as if everybody is connected to everyone through some kind of business deal, though by far not every deal comes to a successful end. It is hard to identify an overall pattern or a clear economic calculation behind the practices based on thorough information on a wider market. Instead, it can rather be characterised by short run opportunism or, as Geertz has put it, 'a carpe diem attitude toward commerce' (Geertz 1963: 36). What has not been stressed as strongly in the work of Geertz, however, is the urge of the traders for mobility. Either following people or picking up and distributing objects, Matar and Seif are always on the move (*kuranda*) from and to the warehouse, from and to other traders, potential buyers, and other people who might be able to help. In this respect, it is important that the goods are generally unbulky and portable to be carried easily to potential buyers and back. And although bigger goods like a refrigerator cannot be carried around, they at least have to be easily storable and long-living as one never knows how long it will take to find a customer. This is generally done through passing by at a number of friends in different places of the city in order to maintain one's connections and exchange news, which among other things always also concern goods that are currently available. Hence, the traders never sit anywhere for very long, always driven to look for new deals. A lot of the arrangements are made on the road when sitting in a car or on a motorbike, or a Vespa, together, or when walking through the streets of Kariakoo. Due to heavy traffic on the major roads the daytime mobility is mainly restricted to the areas near the city centre, whereas longer tours to see someone further away usually take place later in the evening. And even if a lot of this mobility does not bring any direct trading success – as several days or sometimes even weeks can pass without having sold anything – there is always a chance that one might sell something. And even if not, it at least clearly expresses one's constant busyness.

Looking at Matar's travel diary over the last months, it also becomes clear that this mobility does not only extend over the city of Dar es Salaam, but that he still visits other places – Zanzibar and Pemba – to refresh his connections, look what is going on and if there are any business opportunities over there. Apart from briefly visiting his relatives, he clearly spends most of the time with his friends, accompanying them on their movements through Zanzibar and Pemba, sitting in a shop or – in the evening – on a verandah (*baraza*) with them. In Zanzibar, he usually only stays for a couple of days, sleeping either at his cousin's place in Vuga or in one of the 'ghettos' of his friends. Being in Pemba, he usually stays with his parents and younger brothers in Wete. Having lifelong, close connections to a lot of people there, Wete is the place where it is most easy for him to live without having to spend money. There are enough households where he can simply turn up at lunchtime and will surely be offered to eat with them. Moreover, knowing many of the shop owners in town, he also always knows a place where he can get something he needs without having to pay immediately. Nevertheless, neither wanting to overstrain these relationships nor wanting to give the impression of having nothing to do in Dar es Salaam – and especially older people do not hide their critical comments lamenting the laziness of the young generation – after a while he is also happy to travel back to Dar es Salaam, where he can live more independently, more or less hidden from the eyes of critical relatives and neighbours.

…*kushughulikiwa, kutumwa* – …being looked after, being sent

This example of Matar shows how mobility – be it directly related to the trade of specific goods or to look for potential clientele – is also strongly influenced by the kind of relations between the different people involved. Whereas it is generally expected to regularly visit one's family and report on one's situation – even if only very briefly and superficially – it is especially this kind of family and also family-like connections that are most sturdy and therefore play a crucial role in allowing for a live without (regular) income. Seif and Matar also maintain and have established a number of these relationships in Dar es Salaam, including Seif's uncle, Matar's sister, a cousin and some older friends, whom they can also ask for a private credit (though not always sure of being granted one, depending on the current financial situation of the people they ask). But, apart from providing the young traders with free meals or sometimes even with money, older friends, in particular, like the one working in the Bureau de Change or Matar's host, are also the ones asked for advice and recommendations, and are generally talked to very openly about their current state of affairs.

Besides being looked after (*kushughulikiwa*), these patronage-like relationships also allow the 'patrons' to demand certain services from the young traders. Often, this means that Matar and Seif are sent (*kutumwa*) to pick something up for their 'patrons' or bring something to somebody. This way they become involved in their 'patron's' trading practices by sometimes being given the chance to earn a share. In any way, the 'patrons' generally increase their mobility while at the same time depriving them of a certain choice for or against certain activities. When Seif's uncle calls Seif to do something for him, or when Hamud, Matar's host, tells him to drive to Temeke to pick up some money from a friend, it is simply not possible to refuse in order to engage in anything else more profitable for oneself. In this respect, it is important to have somebody whom they can send and give orders (*kutuma*) to look after their own things. In return, they try to care for them (*kuwashughulikia*) by sorting out a place for them to stay, helping them to arrange some business deals, or – when having the means – they might also give them some money. This is also the kind of relationship that has been developed between Badi and Manju on the one side, and Seif and Matar on the other. And overall, it is also these kinds of relationships that, by tying together a multitude of people and things, lead to and constantly reproduce a complex meshwork in which young traders are always on the move, often without a clear purpose or long-term success, but generally convinced of wanting or rather having to engage in trading activities.

As Geertz points out, 'the bazaar-trader is unable actively to search out and create new sources of profit, he can only grasp occasions for gain as they fitfully and, from his point of view, spontaneously arise' (Geertz 1963: 29). In the perspective of Badi, Manju, Seif and Matar, the central criteria for success is to be persistent, to keep going here and there to look for opportunities (*kuhangaika*), thus making mobility instead of economic calculation the major marker. And indeed, regarding the physical mobility of the young traders one can clearly state that this is a more or less permanent feature of their everyday lives. Its steadiness – if necessary or staged – does not only make it become something very mundane, but also serves as a way to conform to the expectations of the wider social context by expressing one's busyness. However, this only holds true for the physical mobility. With regard to mobility in a more metaphorical sense of the term, it becomes clear that these traders are rather immobile. Instead of making progress, being able to substantially expand their trading activities or trying out something different, they conservatively constitute and constantly reconstitute their connections, sticking them and sticking to them by engaging in the same practices in the same way again and again. Overall, this way of dealing with goods that have travelled long distances is mainly characterised by a rather static way of life of the traders.

Coming back to the origin of the goods traded by Badi, Manju, Matar and Seif, the question on the effects of the containers full of used electronic goods sent from a cousin in the UK can now be answered in a more sophisticated way. Whereas, in general, the opinion has become popular that migration can create and support sustainable development in the places of the migrant's origin, due to their personal connections and knowledge of the community's problems and needs, this example shows that this does not always have to be the case. Though it is possible to acknowledge Badi and Manju's cousin in Southampton the ideal of participation by choosing to support the preferred activity of his cousins and letting them decide how to proceed from the time the container is on its way, the lack of control (even in this familial context) and the low return rate concerning the money agreed upon have made this kind of trade unsustainable for him. After the sending of the 4th container of goods he was thus forced to give up or at least make a long pause. In effect, investing in container trade, from the perspective of the cousin in the UK, has neither led to considerable improvements or more financial stability on the side of his cousins in Dar es Salaam nor has it brought any profit for himself. But, even though, the economic success of this kind of investment is often marginal and, in many cases I observed, they even resulted in a considerable loss, they still contribute to the sense of a translocal space. Inducing this kind of trade not only particularly supports and emphasises existing practices instead of changing patterns and connections, it makes the cousin based in the UK become a part of translocal trading practices too. That the importance of this goes far beyond any economic calculation will be illustrated in the following.

TRADING WITH MOBILE PHONES: TRADING CULTURE BEYOND TRADERS

'Petty commerce provides for the trader the permanent backdrop against which almost all his activities occur.' (Geertz 1963: 30)

When carefully following particular objects on their way through the hands of different people, it becomes clear that the aura of trading practices extends far beyond the explicit trading sphere deep into ordinary life. Apart from focusing on the buying and selling of goods, gaining a deeper insight into the life course of an object offers a more nuanced picture regarding different ways of relating to, using, disposing of and acquiring things, exemplifying creative and modulated ways of living an 'trader's way of live'.

As already indicated with regard to the trade journey through the Tanzanian interior, mobile phones are a primary material form in contemporary Swahili trading practices, both as objects of exchange and as technologies of communication. They are of vital importance not only as a trading good but also for the organisation of trading practices. Furthermore, mobile phones have come to play a crucial role in everyday life more generally and can be regarded as a great facilitator in the constitution and maintenance of translocal connections. Nevertheless, mainly due to the costs, mobile phones are far more often used to communicate over short distances strengthening and expanding ties to people nearby. This is also the case when looking at the mobile phone Saada got from a young Canadian woman who was staying at her place while doing voluntary work in a primary school in Zanzibar.

After having received the phone, the *Siemens CF62* is first of all made 'Swahili' by completely relating its tasks to an understanding of Swahili culture. The phone's ringtone is immediately changed to one of the famous and fashionable Arabic songs to express its owner's personal style, clearly indicating Saada's translocal connections beyond Africa. Overall, Saada spends the equivalent of US$1 a day on the phone, usually sending (*kutuma*) one of the neighbouring children or the nephew of her husband to go to the shop nearby to get her the credit. Sometimes she also calls her husband who sells top-up-vouchers in his internet-café to tell him to buy one for her, which she mainly uses to communicate with her female friends and relatives. Although this sort of communication has existed before the widespread use of mobile phones, mobile phones facilitate the exchange of news between women through allowing for communication to take place in different spaces. For Swahili women who usually need to have a good reason to leave the house, the phone allows for easy communication with others while each may remain in her own private space. The latest gossip is delivered by phone and with it too the news that Saada got a new phone. Despite neither its appearance nor its features being particularly attractive to her, it is its newness that is appealing. When leaving the house to do some shopping, the phone is carried in her handbag, sometimes even just held in her hand so that people can see it. Regularly, when meeting some friends, they make a remark about her new phone, either admiring or expressing a certain jealousy that she is able to get new mobile phones in rather short intervals. Her previous phone she had only acquired less than a year ago.

Borrowing and lending

Over the next weeks, the phone is regularly lent to friends and theirs are borrowed in return. By practicing this kind of material exchange, the phone moves between different users, thus preserving its newness. As Saada states, it is completely undesirable to have the same phone for a long time. Instead, it is important to express the connectedness that allows her to arrange this kind of deal, showing that she does not need to stand still in life. Changing mobile phones can therefore be regarded as a way of avoiding stability and ordinariness. The mobile phone becomes an expression of an imagined adaptability and part of striving for change and progress. Hence, without actually selling or buying anything, these practices of lending and borrowing can be interpreted as a clear reference to the ideal of trading, which also includes a high circulation of goods as an indicator of success.

Nevertheless, in order to find others who want to take her phone in exchange, it is important that the phone itself is desirable enough, i.e. seen to enhance the social status of its user.

With the exchange between Saada and her friends particularly consisting of the sending and receiving of wedding pictures and religious picture messages, it is perceived as a deficiency that this phone is not able to take and send pictures. Not being a model that attracts any special attention from her friends, Saada therefore soon decides to sell it together with her old phone to be able to buy the long-desired Motorola flip phone, preferably in pink. As Katz and Sugiyama (2006) emphasise, the acquisition, use and replacement of a mobile phone can often be related to its fashion attentiveness. This surely has to be seen alongside Saada's financial capabilities and, in this case, it is the additional selling of the 'unfashionable phone' that allows for a better expression of her sense of self through a phone that is seen to more accurately represent her cultural ideal. Here, despite its perceived deficiencies, it is particularly its exchange value that gives this phone its power. To succeed and realise a 'better' phone by selling her old one, Saada asks her younger brother Nassir to help her. He is known for trading mobile phones and is given the phone to arrange a deal.

Keeping objects in-between different owners

Nassir, similar to Badi, Manju, Seif and Matar, decided to engage in trading practices in order to earn a living when, after failing the exam to enter Form V (12th grade), there seemed nothing else to do. Inspired from Saada's husband, his brother in law, who – apart from having the internet-café – also runs a shop in Mchangani, he first opened a shop for clothes in his hometown Wete in the north of Pemba. But without selling as much as needed to cover the costs, after two years he finally had to give up the shop. While thinking of what else to do, still hoping to be taken to Canada by his oldest brother, he started to trade in mobile phones. To him, a mobile phone is a bulk commodity, omnipresent and never failing to be in need of a new owner. Therefore, his task is to create new owners by tempting people to buy and sell, with mobile phones always being at the centre of the exchange process.

Receiving the phone from his sister, it is immediately included in a network of trading practices and added to a pool of mobile phones that are for sale. With one of his SIM cards entered into the phone, it now becomes connected to a number of potential buyers and people who might be able to find one. Apart from extending its connections, the phone also becomes more mobile, as it is generally carried around in his trouser pockets. Not having a shop, Nassir is generally on his way to see and inform himself of potential buyers or sellers. Though still spending most of his time in Pemba, he tries to travel to Zanzibar almost once a month and also goes to Dar es Salaam as often as possible to look for business opportunities, orders and possible customers. Moreover, his job involves following people to get and deliver money that is often paid in instalments. This busyness and mobility can be seen as a type of resistance to the routine of everyday life. Especially in Pemba, clearly perceived as an economic periphery, engaging in small-scale trade is considered one of the few alternatives to hanging around at a *maskani* (meeting place for a group of friends) the whole day. In this context, in which Nassir and his friends regularly express how they feel stuck and meaningless, small-scale trade in mobile phones becomes both an occupation and an opportunity.

Fig. 37: Meeting potential customers in Pemba

Moreover, although Nassir cannot afford to buy an expensive phone, his job enables him to use and be seen with the 'best' phones that come into his hands. From his point of view, this temporary enjoyment contributes to his reputation as a trader and raises interest in his goods in general. In this respect, when having an attractive phone in his hands, it is sometimes more attractive to keep it for a while, instead of having to give it up through selling it. Trading highly demanded goods always also means to be able to use these highly demanded and culturally valued goods. In effect, even when sometimes not having 1 Tanzanian Shilling at one's disposal, trading with and therefore also commanding attractive goods still secures a certain social status.

Among the phones used by Nassir, Saada's one takes up a middle position. Neither being cheap and therefore attractive to people with little financial means nor being one of the latest models to attract people who can afford more, makes it less appealing to potential customers. Along with other phones, it accompanies Nassir who usually carries two or three phones with him to show to any potential customers he might meet. The phone's ability to travel well is therefore important both for the user and the mobile trader as it can be carried easily and allows for spur-of-the-moment business and exchange on the move (cf. Bender 2004). However, many phones in the hands of a young person might lead to accusations of theft and as it is complicated to prove one's authority to trade, the phone is sometimes left behind to wait its turn. Occasionally, it may be left with someone who has a phone of similar value to sell which Nassir is then able to borrow in order to present it to a potential customer. Over the course of these different trading activities, nobody is ever left without a phone. Depending on its attractiveness, the phone is taken and shown to certain people, or left behind with other people to be able to present another model instead. It soon becomes obvious that, among the things contributing to a successful deal, the mobile phone itself is one of the most indispensable and essential. It is made a central object of the communication between the people involved and becomes an active participant in the trading network. In the phase when the phone moves from one owner to the next, it does so by passing through the hands of numerous users, showing again the multitude of relations involved in any transaction. Instead of a one-phone-one-user relationship, the phone is part of and embedded in complex connections that complement each other but always remain mobile and changing.

After two months, Nassir eventually benefits from these connections and finds someone in Pemba who buys the phone at an acceptable price. Although this phone may not have generated a huge profit, in his opinion it has still contributed to his new career as a trader and to asserting himself in the traders' world. Nevertheless, sometimes it also happens that he finally returns a phone to its original owner, giving up the endeavour to sell it. But even if he does not manage to sell a phone in months, carrying a number of them around and looking for customers still gives him the air of being a trader.

Connecting to multiple settings

The phone has now been bought by Rashid, a young man from Pemba, for whom this is his first mobile phone. As he argues, in his hometown of Wete, his everyday life has been characterised by the close proximity between his home, family, workplace and the *maskani* where he meets his friends so that a mobile phone had not seemed important. People knew where to find him so he preferred to save money instead of spending it on 'superfluous' additional communication. In contrast to most of his friends, he does not yet share the view that not having a phone marked an individual as particularly lacking but he admits to a recent desire to own one himself, instead of having to use his friends' phones. When he decided to leave his job for a while – he has worked as a barber – to go to Dar es Salaam and look for new trading opportunities there, he also decided to buy a phone. He is convinced that he will need to contact people in Dar es Salaam without knowing where to find them and he therefore chose this simple *Siemens CF62* that would enable him to connect to people in an unfamiliar environment and at least at the moment still characterized by distances rather than proximity. This phone is therefore meant to help him to cope with these new distances.

In Dar es Salaam, Rashid is found via the phone, and it is primarily the names and numbers saved on the phone that belong to people he knows or has heard of that help him to contact and meet people. The phone has become a memory preserving and reminding Rashid of useful connections and communications. This way, he also gets into touch with Said, who lives in the 'ghetto' with Seif, whom he also knows as a close friend of Nassir. They soon agree that Rashid – while in Dar es Salaam – can stay in a room with either Seif and his brother or Said and Hilal. From this place it is not far to the shop of his friend Nassor where he spends most of his time. When people he knows have something to sell they contact him, telling him to find a customer. When he manages to sell the good for a better price than anticipated, he can keep the surplus, paying for his daily expenses in Dar es Salaam and hopefully also soon allowing him to save some money. When he is sitting in the shop with his friend and some others, he usually puts his phone on the desk in front of him, awaiting its next incoming call or text message. As it does not have the ability to play music in mp3-format, it is rather his friends' phones that are used to listen to the current East African music charts, either aloud or by sharing the headphones. He is happy however that his phone at least gives the impression of having been in use for many years by having some scratches and looking more hackneyed. Its appearance makes it easy for him to appear as somebody who has had a phone for a long time. And especially as his friend's shop is specialised in mobile phones and accessories, it covers his own inexperience in the field. And at night when tariffs are lower, Rashid uses the phone to catch up with friends or talk to girls without being disturbed. TIGO, these days, has a special tariff called 'longer longa time (*longa* means chatting) in which from 11pm–5am one minute only costs 1TSh, the equivalent of 0,001€) These conversations emphasise the importance of establishing and confirming interpersonal relationships through talk and, as already observed by Abrahams (1983), pursuing friendships is often more important than just passing on a message. After the last conversation is finished, the phone usually gets recharged over night. During the day it would be only reluctantly left behind in the 'ghetto', as you never know who might pass by and take it.

It is generally felt that the phone plays a crucial role for young men in adapting to and leading a 'bongo life' (*Bongo*, meaning brain, is a common nickname for Dar es Salaam and sometimes even for the whole Tanzanian mainland). For these young men in particular, who are always on the go, it makes them feel reachable while on the move and allows them to be part of a mobile society. In the case of Rashid, the phone definitely plays a crucial role in finding his way around in Dar es Salaam and helps him to strengthen existing as well as build up new relations to people. Furthermore, many of his arranged deals could not take place in the same way without being able to arrange meetings on the phone. Nevertheless, after three months in which he has not been able to save any money despite having been constantly on the lookout for trading opportunities, he realises that moving to Dar es Salaam has not got him very far regarding his business plans.

He therefore decides to go back to Pemba and sell his phone to a friend's shop before he leaves. This allows him to raise cash for the ferry and to buy some goods to sell in Wete, so that he will definitely remain a trader.

In Aggrey, one of the busy streets of Kariakoo, about a hundred shops sell mobile phones and items connected to their usage and appropriation, such as SIM cards, top-up vouchers, chargers, shells etc. Although most of the shops only consist of a few square metres, they all have a glass cabinet at the front displaying their selection of mobile phones – different brands in different rows, arranged by price or by the time they have been on display. This road is full of mobile-phone shops offering a broad (if not complete) range of models that are available in stores all round the world today, from the latest Sony Ericsson design to the Motorola flip phone in all different colours and some of the old and simple ones by Nokia and Siemens. Looking at this collection, it becomes obvious that thousands of processes of acquisition and exchange are involved, stretching across space and time. But what is less clear is if the high mobility of the objects and the high physical mobility of the traders also lead to mobility in the sense of change and progress. Despite always being on the move and arranging new deals with a multitude of people, their way of life still appears to be rather stable and stagnating. However, although only a minority seems to make a good and regular income out of these trading practices, this way of life proves to be extremely persistent as an ideology, with practices of borrowing and lending, as well as keeping things in-between, allowing to perform as traders, even if – as in the case of Saada and Rashid – that is not their (only) profession.

Fig. 38: Mobile phone shopping centre in Aggrey, Kariakoo

AN IDEOLOGY OF TRADE: REFLECTIONS ON CULTURE AND ECONOMY (III)

'No denial of free will is implied, nor is the scope for individual achievement or resourcefulness belittled. It is simply that we are all players in a great profusion of games and that in each cultural arena the entire team, knowingly or not, follows the local set of rules, at most bending them slightly. Only a half-wit or a fool would openly flout them. But as in chess, the possibilities for creativity and modulation are virtually infinite.' (Zelinsky 1973: 70–71)

Following the flow of goods, if in the context of small scale trading activities or in more ordinary settings, illustrates well how the different practices of exchange are 'at once an economic institution and a way of life' (Geertz 1963: 30). Whereas it would be difficult to understand the wannabe traders' behaviour and especially its persistence from a solely rationalistic economic point of view, it shows how important it is to rather carefully explore the meaning trading practices have in respect to what might be labelled 'Swahili culture'.

The meaning of the term culture has changed considerably over the last hundreds of years, turning from a term indicating 'cultivation or tending', via 'civilisation' to a complex concept with an ever extending usage referring to the opposite of nature, a total way of life, distinctive markers to differentiate groups of people, as well as to its representations such as texts, architecture, music etc. (cf. Williams 1976, see also Mitchell 2000: 13–16). As it is neither possible nor reasonable here to give a full account of the history of 'culture', what I want to focus on instead is how reflections on 'culture' can contribute to an understanding of how it comes that great numbers of young (and also older) Swahili engage in small-scale trading practices, grapple with this way of life and stick to it, although economically-speaking they often fail and hardly manage to become established traders. In doing so, I restrict myself to broader strands which have been dominant in different phases in the discipline of cultural geography.

The engagement with culture among geographers, mainly derived from a strong criticism to environmental determinism and social Darwinism developing in the late 1890s and the beginning of the 20th century, led to a position in favour of historical particularism and cultural relativism as advocated most prominently by Boas, the so-called founding father of cultural anthropology in the United States (cf. Stocking 1982). Clearly informed by these ideas as well as by Herder's critique of instrumentalism, it was Carl Sauer who steered cultural geography away from environmental determinism through placing culture right at the core of the discipline, though mainly remaining concerned with the material landscape as a manifestation and outcome of culture (cf. Sauer 1925). Culture in this respect is rather vaguely seen as historically, geographically derived difference, appearing to be somehow larger and greater than the people who compose it (cf. Mitchell 2000: 24). One of Sauer's students, Zelinsky, further developed this idea to theorise culture as 'superorganic', regarding culture as 'something both of *and beyond* the participating members. Its totality is palpably greater than the sum of its parts, for it is superorganic and superindividual in nature, an entity with a structure, set of processes, and momentum of its own, though clearly not untouched by historical events and socio-economic conditions' (Zelinsky 1973: 40–41). This idea of culture as having a life of its own and determining the lives of the people who are part of it, independent of their will or intention, has been met with harsh criticism from the side of geographers who wanted to invent a more nuanced, more 'sociological' sense of culture, which they thought could allow for a better understanding of conflicts, questions of power and the way culture worked in society (cf. Duncan 1980, Jackson 1980, Cosgrove 1983). As a result, some geographers suggested reducing culture to the interaction between people concentrating on the 'many problematic social, political, and economic relationships that govern our lives' (Duncan 1980: 198). This position then met with the theoretical developments concerning 'culture' that have taken place in the British context of Cultural Studies in the early 1980es, at that time taking off and entering cultural geography by 'implying a need to focus on the role of space and place in adjudicating cultural power' (Mitchell 2000: 58). Bringing views of culture as inherently social and political into 'creative tension with the humanism that had emerged within geography in the late 1970s with its interest in issues of space, place and meaning' is what Cresswell calls the 'genius of the 'new' cultural geography (Cresswell 2010: 171).

In the early 1990s, cultural geography became extremely influential in Anglophone geography, strongly informing other subdisciplines such as historical geography, political geography and economic geography. As highlighted by Cresswell (2010), it was particularly the field of social geography that became closely tied to cultural geography, expressed for example in the journal title Social & Cultural Geography with its first issue published in 2000 (cf. Philo 1991). While some geographers feel their identity as social geographers to be threatened by the increasing influence of cultural geography, others argue for the opposite direction by stating that the close link between cultural and social geography has made cultural geography far too social (cf. Cresswell 2010: 171–172).

Whereas understandings of culture have been vividly discussed in Anglophone cultural geography, especially in the wake of what has been termed 'new cultural geography' in the 1990s (cf. Mitchell 1995, Jackson 1996, Duncan & Duncan 1996, see also Cresswell 2010 and Crang 2010 for recent reflections on the issue), in Germany, a direct and more sophisticated engagement with the meaning(s) of culture so far remains rare (cf. Popp 1993, see also Boeckler 2005: 28 for the same observation). This might have much to do with the fact that in Anglophone geography, a considerable number of cultural geographers clearly position themselves in the humanities and reject 'the social as a realm that provides explanation' (Cresswell 2010: 172), while in the German-speaking context this is far less the case. This is also reflected in the edited volume *Kulturgeographie,* published in Germany by Gebhardt et al. in 2003, in which the editors state that cultural geography is not understood as a 'culturalistic' geography, or as a geography that tries to replace the social orientation with a cultural one, but more generally as a constructivist perspective (Gebhardt et al. 2003, see also Blotevogel 2003, Redepenning 2007). Instead of an idiosyncratic cultural interest and a search for meaning, it is rather an increasing attention to constructivism and contingency that lies at the basis of the discussions under the heading of *Neue Kulturgeographie* in Germany over the last decade (cf. Lossau 2008).

However, as part of a postmodern critique of the 'modernist' concept in which culture has been 'deployed to stop flux in its tracks and to pull out the essences of a situation, creating the contours of a stable "way of life"' (Mitchell 2000: 75), following the 'cultural turn' geographers in both contexts have been concerned to overcome this idea of culture as an ordering device by instead pointing to disorder, ambivalence and fuzzyness. According to Boeckler, it would even be adequate to call the 'cultural turn' a fight *against* culture, which resulted in deconstructing the concept, redefining the term in favour of a heterogeneous multiplicity, and extremely expanding its usage (Boeckler 2005: 31). It is particularly the essentialising and homogenising tendency of previous understandings of culture that has been and still is harshly criticised from a more postmodern and poststructuralist point of view. As a result, culture has become everything, superfluous, fluid, or politics; and although the number of work under the heading of 'cultural geography' has increased rapidly in the Anglophone and subsequently also in the German-speaking context, it is still strikingly rare for empirical studies to actually reflect on any of these myriad and vague definitions of culture.

But is it really sufficient to refer to culture as contingent, constructed, processual and negotiable when striving to account for the context presented above? How can such an understanding come to terms with the fact that trade has been so persistent among Swahili despite economic disappointment and remains a somehow taken for granted practice, a given? When asked about the motivations and reasons for engaging in trading practices, the most frequent answer is a reference to 'Swahili culture'. Culture here is clearly understood as 'the proper' Swahili way of life, encompassing language, food, dress, religion, habits and values – values, attitudes, ideas and opinions on an ideal way of life, including economic, political, religious and social activities. 'Swahili culture' is somehow there, as a self-understanding, an ideology, having considerable power in itself. Thus, despite the understandable and convincing criticism of essentialising concepts of culture, this context illustrates how essences are constructed and play an important role in empirical situations. 'Culture' somehow imposes itself on any interpretation of the everyday life of the traders presented above. Not as a force to determine their lives but rather as a kind of 'self-service store', offering a set of 'ideal' practices to be followed (while others are not available). 'Swahili culture', in this respect, can serve as an excuse, reasoning and legitimisation, also in times of stagnation or economic failure, by giving cultural valorisation to trading practices. As Mitchell states, it is therefore crucial to look at the ways in which people make sense of what they are doing, how 'activities are reified *as* culture' (Mitchell 2000: 77), taking into account however that the engagement with culture is always also culturally embedded and not necessarily strategic or rational as generally assumed by those that reduce culture to politics.

Following a relational perspective, 'Swahili culture' can be understood as a punctualisation of a multitude of different intermingling branches and flows, a simplification, which has gained a considerable durability and efficacy. In the following, I will therefore elaborate on at least two of these strands that are frequently referred to by the traders in serving as a reason for their engagement in trading practices.

A glorified past: Social and cultural prosperity through trade

With trade having been the central driving force in the establishment and maintenance of translocal connections, which play such an important role in everyday life until today, trading practices are put right at the core of 'Swahili culture'. As Sheriff (2009: 173) has put it, 'one of the primary processes of interaction across the Indian Ocean was trade, which was the *raison d'être* for much of the communication' between different places along the Indian Ocean coastline. Nevertheless, from the beginning this 'mundane and uncelebrated peaceful trade' (Sheriff 2009: 174) is said to have been interlinked with intermarriages, social and cultural exchange (cf. Casson 1989: 61). Although the movements of goods surely have been the primary reason for establishing translocal connections, the related movement of people may have had an even higher impact with respect to 'Swahili culture'. Since the emergence of those who are now commonly known as 'the Swahili', generally dated back to around the year 800, the mercantile character of Swahili culture has been emphasised (cf. Nurse & Spear 1985, Horton & Middleton 2000, Kresse 2007). This is pointed out as well by Middleton, who states that 'the Swahili role of merchant between distant and "foreign" traders has been central to Swahili culture, even though only the patrician minority were engaged in the actual exchange' (Middleton 2004: 79). These patricians (*waungwana, wastaarabu*) were seen to be staying at the top of a clearly stratified society, distancing themselves from the *washenzi* (barbarians) – or less drastically put as *wenjeji* (here translated as commoners) by Horton and Middleton (2000) – with the majority of Swahili nevertheless being positioned somewhere on the 'gradient between [these] two idealised poles' (Nurse & Spear 1985: 25). What is important here in respect to the understanding of 'culture' is that, whereas *ushenzi* is considered as 'wilderness" and 'unculturedness', *uungwana* is generally translated as 'civilisation', hinting at the idea of high culture that can be possessed or not (cf. Baumann 1999, Elias 1976). Moreover, being a patrician or – as they are also called – nobleman, generally implies to have gained one's wealth through the engagement in trade (cf. Kresse 2007: 52). This intermingling between social status and occupation is also well expressed in the Kiswahili term *tajiri*, derived from the Arabic *tāǧer* (translated as 'trader'), which refers to a rich person or someone who engages in commerce. To be a rich person (*mtajiri*) is not only related to having money but also refers to being rich in social status. In effect, besides economic supremacy traders also enjoyed political power and, together with religious leaders and learned men they were dominating all aspects of life (Le Cour Grandmaison 1989: 179).

Apart from the discovery of oil in the Persian Gulf in the beginning and middle of the 20th century, which led to alternative sources of income and a complete restructuring of the economies of some of the countries decisively involved in the dhow trade, it is particularly the Zanzibar Revolution in 1964 and the politics of exclusion that followed this event that have resulted in an immediate decline of translocal trade to and from Zanzibar and disrupted the long-distance dhow trade more widely. Nevertheless, as soon as the trade restrictions were officially loosened in the mid-1980es (and probably even before), translocal trading practices were taken up and have increased constantly since then.

Although the glorious and economically most successful times of Zanzibar have long passed, as the example of Badi, Manju, Seif and Matar presented above shows, 'at the small-scale, the personal exchange system of Swahili merchants continues' (Middleton 2004: 87). In contrast to some Swahili scholars who mainly point at the differences and paint a very black and white image of Swahili trade in the past (seen as 'precapitalist' and 'precolonial', based on kinship, friendship and trust) and more recent trade (as 'capitalist' and '(post)colonial', characterised by impersonal relationships and a high dependency on (communication) technologies), I thus consider it to be more fruitful to also look at the continuities. When it is said, for example, that in the past 'these distant traders rarely, if ever, came into direct contact but dealt indirectly with one another through the Swahili merchants of the coast' (Middleton 2004: 79), examining the trading practices of Badi and Manju shows that this is still what they do. Furthermore, even though communication technologies do indeed play an important role in the organisation of the trade today, social connections based on relation or common neighbourhoods do so too. Finally, as the representation of the young traders' everyday trading practices has illustrated, it is not that rationalistic concerning the maximisation of their profits either. Until today, the general idea is that to be culturally and economically successful, one has to engage in commerce, and that those who engage in commerce also become rich, not only financially but also by moving up the social hierarchy. Almost all the young people I have spoken to claim that, whatever they might do to earn their living, they would also definitely engage in some kind of trading activity, and it is still common to hear that as a Swahili you *have* to be a trader. In this respect, trading practices are still recognized as a worthwhile activity, not only for the anticipated profit, but also due to the historical role of trade in the Swahili context that makes it a central part of the understanding of 'Swahili culture' today. Particularly in a context in which many mourn the good old times – with the older generation regularly referring to the young ones to illustrate how 'Swahili culture' is loosing some of its central values (cf. Saleh 2004) – engaging in trade can be seen as an attempt to hold up these values and habits, trying to connect with the glorious history of the Swahili. Being part of translocal trading connections gives them the air of the cosmopolitanism and maritime ethos often asserted to 'Swahili culture', differentiating them and making them 'superior' to 'the Africans' (see *Facing 'Africa': Views from the coast* on p. 81, regarding the persistence of this idea with regard to the relationship between 'Swahili' and 'African'). Moreover, and less pretentious, engaging in trade (*kufanya biashara*) in the way described above, at least indicates a certain busyness and effort, dismissing possible accusations of laziness (*uvivu*), vagrancy and immorality (*uhuni*).

An unjust present: Trade as a bypass to political discrimination

Concerning this second point, it is less the cultural history than the political present that serves as a welcome argument to justify and valorise one's engagement in trading practices. Whereas in the previous paragraphs I have shown how trade is – and has always been – regarded as an ideal way of life, in the following, I will illustrate how trade can be seen as ideal way of life because there simply appears to be no alternative.

A brief excursion into the political present of Zanzibar

It was in 1890 with the declaration of a protectorate that the British formally seized power in Zanzibar, which had since the 1840s been the capital of the Omani Empire in East Africa as well as the seat of the Sultan. Supporting the social and political dominance of the so-called *waungwana,* mainly consisting of Swahili with Omani descent, it is the following phase of British colonial rule that is generally seen to have fostered a raising awareness of ethnic divisions among the population, and only more recently, scholars have also started to reconsider the role of Zanzibar's intelligentsia in this racialisation of political thought (cf. Glassman 2004, Larsen 2008: 28).

Soon after independence, these political tensions culminated in an 'anti-Arab revolution' led by the Ugandan John Okello with a force of mainland Africans and members of the Afro-Shirazi Party (cf. Okello 1971, Lofchie 1965). Though accounts of the revolution, its organisation, legitimisation, and support differ, the overall aim clearly was 'to rid Zanzibar of its Arab sultan, to end the economic privileges of Asian and Arab Zanzibari, and to impose a socialist government run by Africans and oriented more towards the coastal states of East Africa than to lands beyond the sea' (Gilbert 2007: 169). Whereas in the eyes of members of the Afro-Shirazi-Party it was a 'glorious overthrow of a slave-holding feudal regime by an oppressed African majority', for many other ('Arab-oriented') Swahili this meant a horrifying experience with about 6000–10000 of them killed and many deported or dispossessed (Cameron 2004, see also Clayton 1981, Gilbert 2007). Only three months after the 'revolution', in April 1964, the union between the Revolutionary Government of Zanzibar headed by the Afro-Shirazi Party (ASP) and the Tanganyikan Government was formed, creating the Tanzanian nation-state led by president Julius Nyerere. Until today, Zanzibar has remained a semi-autonomous part of the United Republic of Tanzania, with its own government, but without an independent foreign, security or monetary policy, an agreement that is critically debated – particularly on the Zanzibar side – until today (cf. Larsen 2008: 29).

Following a socialist ideology, already in 1970 most of the urban property in Zanzibar had been nationalised and the government had taken a politically isolationist course, prohibiting most translocal connections, especially the ones to Arabia (cf. Gilbert 2004). An import monopoly was given to the state firm Bizanje (Larsen 2008: 39–40) Moreover, a politics of 'Africanisation' was initiated, strongly marginalising those who had previously been categorised as 'Arab' or 'Asian", implementing among other things the 'Forced Marriage Act' in 1970 which, for the duration of two years, served as a means to eliminate or at least diminish ethnic differences. The 'Forced Marriage Act' made it legal for men of African origin to marry women of Arabian or Asian descent without the consent of the women themselves or their families (cf. Larsen 2008: 29). In 1971, all non-African businessmen were denied to renew their licences, so that all owners of private businesses were forced to take an African partner.

More than 20 years after the revolution, in the mid-1980s, Zanzibar was near a financial breakdown due both to a number of big projects that had failed and the rapid decrease of world market prices for cloves. As a result, the government was more or less forced to start a process of economic liberalisation while also slowly reintroducing a multi-party system (Gilbert 2007: 173). Nevertheless, the new political setting headed by the ruling *Chama Cha Mapinduzi* (CCM, Revolutionary Party, the result of the merging of the ASP and the Tanganyika African National Union (TANU) that took place in 1977) is still met with considerable criticism by many Zanzibari who regard their reforms as an unwanted interference in Zanzibar politics, social structure and culture more generally.

Until today many Zanzibari still feel to be treated unequally, oppressed and deliberately isolated, blaming the government for failed health and education services, poor producer prices for cloves, inflation, high unemployment and a low accountability of its institutions more generally (cf. Cameron 2004: 115). This also manifests itself in the raise in votes for the opposition party *Civic United Front* (CUF) and the violent upheavals in the context of the multi-party elections in 1995, 2000 and 2005 where the ruling party has always been accused of rigging the votes. Political affiliations are often still discussed as an indicator of an either 'Arab' or 'African' orientation, so that the increasing support of the CUF has been publicly interpreted by the government as a demonstration of the remaining wish of 'the Arabs' to 're-arabise' Zanzibar (cf. Larsen 2008: 30–31). In the wake and immediate aftermath of the 2000 General Elections many people fled from Zanzibar as a result of violent clashes between the two parties that even involved the army shooting into demonstrating masses. Under the increasing pressure of international donors, in October 2001 an agreement (*muafaka*) was signed as a result of reconciliation talks between the two parties formulating aims to reform the Zanzibar Electoral Commission, achieve free and fair elections and also allowing for a possible coalition. After another disappointment in 2005, where the CCM was again announced as the winner and remained in power without any participation of the CUF, finally, in 2010 (after I had done most of my fieldwork), a coalition Government of National Unity was formed. However, the longstanding doubts concerning the reliability of the government are still sensible.

As this brief excursion into the political present of Zanzibar illustrates, the relationship of many Swahili towards the government has for a long time been characterised by the experience of oppression, isolation and violence. According to Larsen (2004: 138) it has especially been in situations of conflict and perceived injustice that people tended to explain their way of life with reference to an essentialising 'African'-'Arab'-divide, in which 'Africans' are clearly associated with the government. On the one hand, this has led to the conviction that as 'Arabs' (or 'Zanzibari') they would not have a chance to be supported by the political regime since they were generally associated with the opposition and are therefore left to look for income opportunities elsewhere. On the other hand, many Swahili state – even after the formation of a coalition – that they do not want to be associated with the government, although it is hard to say how many young Swahili would actually refuse a government position if they were offered one. This leads to a very negative representation of government-paid jobs that are associated with low or irregularly paid wages and a high dependency on the political apparatus. Moreover, another very important aspect is that, in accordance with the translocal orientation of the Swahili, it is not a very welcome thought to affiliate with a government that has tried and is still seen to undermine these translocal connections by representing a Tanzanian unity that does not conform to senses of belonging and identity of many Swahili. Overall, it can be observed that the different discourses of discrimination and unjust politics serve as a way to gain support for one's more or less successful attempts to earn a living without relating to the government in any way.

> Kuajiriwa hana maana; kuajiriwa ni kama kutumwa tu. Yaani, muhimu uwe na biashara yako mwenyewe. (Anwar, Kariakoo, 27.09.2008)
> (To be employed has no meaning; to be employed is like always being under orders. Thus, what is important is to have your own trading business.)

Trade, in this respect, can be interpreted as an expression of political disinvolvement, earning one's livelihood through a practice as independent from the government as possible and explicitly transgressing the national context. Of course, this independence is restricted due to import legislations, taxes etc. However, there is also a strong tendency to adhere to an informal organisation of trade that tries to avoid payments to the state. Being a trader is generally celebrated as working independently, enjoying a certain freedom in structuring one's working day, working at one's own risk and for one's own profit. It might be questionable if trade can be regarded as an efficient form of political resistance, but discursively it seems to be widely understood as such, or at least as a certain withdrawal from current politics, while at the same time emphasising a crucial aspect of one's culture.

Somehow, this negative attitude towards contemporary politics is not only reflected in the denial of attempting to work in state institutions but also regularly serves as an excuse for not striving for any other career in the private sector. The above-mentioned inequality between Zanzibar and the mainland has severe effects on the quality of education, but instead of trying to work exceptionally hard to still make one's way to university, concerning the majority of young Swahili it seems as if 'political discrimination' as a crucial element of 'Swahili culture' has generally been strong enough to account for one's academic failure while paving the way into trading practices.

This does not mean that I do not agree with the political inequalities many Swahili indeed have had to face. What I wanted to show though is how a remaining 'rhetoric of ethnic essentialism' (Cameron 2004) in current political discourses on an individual level can serve as a welcome argument to support one's engagement in trading practices by making it the only possible and therefore ideal way of life as a Swahili. In effect, a considerable number of young and predominantly male Swahili can be observed in their joint struggles, all trying to become traders in very similar ways. In this respect, 'Swahili culture' is a constant point of reference, offering values, habits and opinions that give a strong valorisation to one's everyday life despite (economic) stagnation. As Geertz points out, 'trading for trading's sake [becomes] one of the primary goals of [their lives], [...] an expression of self' (Geertz 1963: 44).

5TH HALT: STICKING (TO) TRADING CONNECTIONS

Looking at the small-scale or peddling trade in the Swahili context shows how strongly these practices are related to understandings of 'Swahili culture'. First of all, it is remarkable how many of the young male Swahili I met were engaging or trying to engage in trading practices and had at least one foot in the trading business. Second, it is also striking how few of them actually manage to earn a regular living from this trade, making it very difficult to explain their endeavours solely from an economic point of view. In contrast to a clear economic rationale, it seems to be far more which is behind the omnipresent appeal of trade.

To many, trade has become an ideology directly associated with the 'proper' Swahili way of life. Whereas on the one hand, trade is seen as a guarantor of economic and social wealth due to its crucial role in making the Swahili coast as important and successful as it was more or less until the mid-20th century, on the other hand, it is also regarded as the only possibility to establish ones livelihood without having to engage with the national government, which is still very negatively perceived by many. Trade, in this respect, can be interpreted as a way to react to discriminatory strategies of the state and resist governmental structures by relying on a way of life that has not only proved to be successful, but also counts as a core factor with regard to Swahili culture and identity. In a certain way, the way 'Swahili culture' is used to justify and explain the widespread engagement in trading practices, while bypassing usual economic calculations, creates an image of culture as a kind of self-service store that offers specific practices and discourses which help to support one's position in this particular context. Referring to the crucial historical role of trade for the emergence of the Swahili, as well as pointing to the perceived injustice and discrimination of the current government, serves as a way to gain support for their way of life as a trader. At the same time, it distracts from other reasons, such as little education, lack of ideas etc, that might have contributed to their current state as occasional or 'wannabe' traders as well. However, these latter reasons often feature prominently in the discourses especially among elderly Swahili, which focus on the loss or at least the degradation of traditional Swahili values due to the laziness and immorality of the young generation. Without wanting to completely deny that such aspects might indeed play a role, making small-scale or peddling trade an easy option for those who are, for whatever reason, not making an effort to establish any other kind of living, I also want to point to another idea here: Instead of giving up traditional Swahili habits and values, sticking to trade shows how relevant these ideas of Swahiliness still are to the younger generation and how it is exactly these old ideals that serve as a reasoning and driving force into the way of life of a trader – and make it so very popular to surround oneself with the aura of trade.

What the lives of the young traders presented here also show is an enormous relevance of physical mobility in respect to the organisation of trade. This physical mobility indeed is not only a steady characteristic of their everyday lives; it also portrays well how mobility can entail to somehow stand still. Mobility in a metaphorical sense, standing for flexibility, progress and change, generally remains a permanent performance in contrast to the steady physical mobility actually characterising their daily life. Despite constant physical mobility, it rather seems as if many of them stay where they are, continuing with the same practices without getting anywhere, like a hamster in a hamster wheel. Mobility here is closely connected to repetitivity and stagnation, supporting a common way of life based on shared discourses and practices that are rather conservative than progressive. Sticking to trade, when taking into account the economically low productivity characterising the situation of most of the traders, rather seems like a vicious circle than bringing progress and change.

While the ideological meaning of trade is reflected in the omnipresence of trading practices, it is also these practices that contribute to the persistence of this ideology. By involving a multitude of relationships, pursuing these kinds of trading practices further strengthens, tightens and expands the traders' connections, making it all the more difficult for them to step outside or distance oneself from this particular cultural context. It thus seems as if many of the young Swahili traders are somehow stuck in this complex entanglement of ideology and practice. Nevertheless, it is particularly because of their close link to 'Swahili culture' that also these small-scale trading practices play an important role in creating a sense of belonging to a wider translocal Swahili space, even if most of the traders only bob up and down in a small part of it.

SEARCHING HOME IN A TRANSLOCAL SPACE: ECONOMIC DIMENSIONS OF TRANSLOCAL CULTURAL PRACTICES

ECONOMIC RATIONALITIES OF CULTURAL PRACTICES AND THEIR MATERIAL EFFECTS: REFLECTIONS OF CULTURE AND ECONOMY (IV)

So far, the central theme that has run through the three previous chapters has been translocality as expressed and lived through trade as an economic practice that is culturally embedded. Different kinds of mobile trading practices were at the core of each of the chapters, all leading to a theoretical discussion on the relation between culture and economy within these practices. The first chapter has illustrated how family visits and trade journeys through the Tanzanian hinterland are closely intertwined and depend on each other. With regard to the professional traders and their trading businesses, in the second chapter, I have pointed out how not only the traders themselves but also consumers form an active part in translocal trading connections by creatively playing with the symbolic meanings of the goods. And by turning the attention to economically less successful traders, the third chapter has examined how it is the cultural ideology of trade that contributes to its persistence, allowing even those traders to feel part of the translocal Swahili space who, despite their constant busyness and mobility, rather seem stuck in these practices. In this respect, one of the central arguments has been to show how trade within the translocal Swahili space cannot be understood by solely relying on economic reasoning. Instead, the empirical insights illustrate well how important cultural aspects are within the organisation and negotiation of the various trading practices, whereas it is often difficult to explain the trading endeavours with clear rationalities and economic calculation. In general, this argument goes in line with the recent debates on 'cultural economy' in which an increasing number of cultural (economic) geographers advocate the deconstruction of the binary between culture and economy in favour of a more cultural understanding of practices previously classified as economic and researched by following an economic-rationalistic approach. Nevertheless, as Hall (1996: 258) has pointed out, an abandonment of a narrowly deterministic economics has often resulted not in an effort to rethink economic relations and their effects, but rather in a 'gigantic and eloquent disavowal. As if, since the economic in the broadest sense, definitely does not, as it was once supposed to do, "determine" the real movement of history "in the last instance", it does not exist at all!' Arguing for a broader, culturally informed perspective in the interpretation of economic practices, rethinking the relation between culture and economy must therefore not lead to a complete dismissal of material aspects and rational thinking, but – besides cultural aspects of economic practices – needs to also take into account the economic dimension inherent in cultural practices. In this chapter, I therefore now want to turn the focus from 'cultural economy' to 'economic culture', looking at translocality as a cultural practice within its economic context.

On the one hand, this entails to look at the strategic thinking and economic considerations influencing cultural practices. Economic reflections and material constraints here act, on one side, as a *cause* for the specific constitution and figurations of translocal cultural practices. On the other hand, cultural practices also encourage different forms of economic and material exchange. Thus, economy and materiality also have to be regarded as an *effect* of translocal cultural practices. And, as these two strands are closely intertwined, they embed cultural practices in complex and heterogeneous material relations.

In order to tease out the economic dimensions of translocal cultural practices, in this chapter, I will now take the organisation of two weddings as a storyline, concentrating on both the economic reflections informing the arrangements of the weddings as well as their material effects. What will soon become clear is how a lot of the material dimensions of marriages have to be considered in respect to the very mobile, translocal settings in which they take place. Where and how to find a marriage partner across distances and changing places? Where to celebrate the wedding when families and friends are spread over many different locations? And how to decide where to live after the wedding? In all these questions, material and mobile affordances are always closely tied together. Looking in more detail at the flows of people, things and ideas involved in marriages will thus not only illustrate how cultural practices are deeply embedded in economic and material aspects, but it will also open the perspective to the material efforts that go into the construction of home in a translocal space more generally.

FINDING SOMEONE WORTH MARRYING

Samir & Aisha

After the official engagement has taken place, Aisha finally has a legitimate reason to visit her fiancé Samir and his family: it is time to discuss and arrange the wedding with them. Although Samir and Aisha have already been in a relationship for about three years, it has always been difficult for them to meet, as Aisha has been living with her father in the western outskirts of London, almost two hours (on public transport) away from East Ham, the place where Samir stayed with his family. Luckily their families are related, so that, together with numerous cousins and other Swahili of their generation living in London, they would regularly meet at birthday parties, weddings, or religious holidays. And, it was also on one of these occasions – a birthday party at the house of Samir's oldest brother – that Aisha and Samir found the chance to talk to each other freely and started their relationship. Birthday parties in particular are favourable environments for flirting and getting to know each other, since young people dominate the celebration, so that there is less control of the parental generation. But finding legitimate reasons to meet up more frequently and spend time together beside these communal events proves to be far more difficult.

As Aisha had fancied Samir for quite a while, she was used to carefully take every opportunity to go to the area where he lived with his family, hoping to, at least, catch a glimpse of him. As East Ham and its neighbouring areas in East London are one of the hubs of Swahili communal activities in London, being the part of the city where a considerable number of Swahili found housing, Aisha was often able to use invitations from her female relatives an friends living in that area as rather innocent reasons to travel there. And since Samir's two half-sisters are related to her through their mother's side, when visiting them she was even able to spend the day in the same house as him. When finally having a relationship with him, she started to encourage her female cousins even more to arrange gatherings, which could serve her as an excuse to come over and, as importantly, give her a reason why her father should pay for her ticket to use the London Underground. Usually, Aisha's father would support her presence at family events, recognising that this is not only expected and shows respect to the respective host, but also because it gives Aisha the opportunity to enjoy the company she was often missing while living alone with her father on the other side of the city (her mother already passed away). Nevertheless, he would also be very protective, generally not wanting his daughter to risk her reputation by being close to a man before marriage. In this respect, Aisha's father soon became suspicious of her trips and over the course of the three years of their relationship Aisha had to go through a number of harsh arguments with her father.

Although Samir belongs to a well-known and influential family in Mombasa, which usually serves as an important criterion when it comes to the selection of marriage partners especially from the perspective of the parents, he has not yet been very successful in establishing his own life. Born and raised in Mombasa as the third of six brothers, it was at the age of sixteen that he and his family moved to London, following one of his aunts and her family. There, before moving into a terraced-house in East Ham, they were first accommodated in a hostel in Bayswater for a whole year, as it was difficult to find a house big enough for all of them. In this context, Samir had a difficult and not very stable start into the new school and did not do very well in his GCSE exams (General Certificate of Secondary Education, equivalent to the German Realschulabschluss). He then started going to College in East Ham, where he was taking a number of different courses, but neither with a strong motivation nor with a clear aim. At the time of the first careful approaches towards a marriage proposal, he was doing a part-time job at Tesco supermarket. Seeing this, Aisha's father and other close relatives, did not immediately approve of her choice and were very reluctant towards the idea of their marriage. They were worried that Samir would not be able to provide their daughter with a comfortable live and tried to convince her to look for someone in a better financial situation instead. Aisha herself, however, went to college only halfheartedly and regularly missed her classes in order to use the time to travel to East London unnoticed by her father. In spite of her family's concerns, she always kept insisting in wanting to get married to Samir until, finally, her father gave in to her stubbornness. And this summer they will actually get married.

Salman & Raya

For Samir's family there are even two weddings to arrange and celebrate this summer. Samir's older brother Salman, as well, has finally got engaged, after having gone through similar problems with his girlfriend before. Having completed Form IV (equivalent to GCSE and Realschulabschluss) in Mombasa, he did not continue his secondary education and, when moving to London at the age of 19, he did not have clear prospects of what he would do there. When one of his uncles found out about an opportunity to work on construction sites in Wales during the week, Salman, sometimes together with one his cousins, joined in to earn some money. Nevertheless, he never really settled in London and often spent more than half of the year in Mombasa, where he lived in the house of his grandfather.

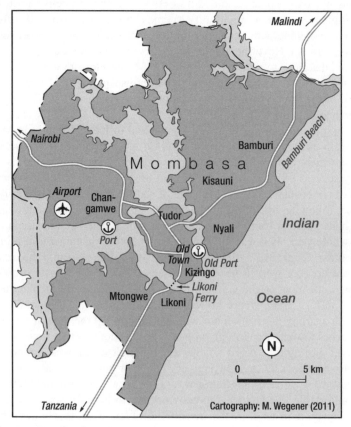

Fig. 39: Map of Mombasa

In 2006, Salman was hired as a player in a football team of Mombasa, so that he stayed there in order to regularly take part in the training as well as to play the games. In addition to that, he tried to increase his income by trading in mobile phones, but not always with the desired outcome.

Salman had met Raya through one of his female cousins who knew her from school in Mombasa. She was living in Kizingo, near his grandfather's office, not far from the city centre. But even though they were living in close proximity they never managed to see each other as often as Samir and Aisha. On the one hand, this was due to the social control in Mombasa that is simply much stronger than it could be in London. Whereas London offers enough space hidden from 'Swahili eyes', in Mombasa there always remains a risk to bump into family members or other acquaintances. On the other hand, there is no direct family relationship between Raya and Salman so that they were not invited to the same family gatherings or celebrations. Their relationship therefore remained to take place more secretly for a long time and was based on far less frequent face-to-face contact.

When Raya finally told her parents that she wanted to get married to Salman, they expressed very similar concerns to those mentioned by the family of his brother's fiancée Aisha. Despite, of course, knowing his grandfather, Salman's own reputation still lacked any educational and economic success and they clearly would have preferred her daughter to get married to someone running a successful business. Nevertheless, they also gave in to her daughter's choice, reminding themselves that he at least comes from a well-known family with strong translocal links. Even before the wedding, first plans were made about how they could live together in London, and especially Raya was looking forward to being able to benefit from the different lifestyle and opportunities in the UK.

Now, after their engagement the two couples can finally communicate more openly and do not have to hide their relationship in front of the parental generation and the elders anymore. However, opportunities to meet each other in private are still rare, not only because they are generally not welcomed before the wedding, but also because of the distance that separates them, especially Salman and Raya at the times when he is in the UK. For this reason, as for so many young couples in this translocal space, virtual meetings still remain crucial in their relationships. Already preceding the official engagement when the four of them had to be even more secretive about their respective relationships, diverse communication technologies had played a decisive role in allowing them to inwardly deepen their relationship.

Love Talk: The material affordances of virtual communication

Throughout the time of their relationship, covering more than three years, Aisha and Samir used the telephone to communicate almost daily, also when her father prevented her from seeing him. Whereas she used the landline to call him when her father was not around or when she could pretend that she was talking to Samir's sisters, Samir usually made an effort to use special mobile-phone-deals that would, for example, offer a flatrate for a couple of days after having topped up a certain sum. This way, they were able to talk for hours, especially in the evening when her father had already gone to bed. At that time, Aisha was able to talk to him unheard by her father, discussing with Samir when she could try to travel to East London again or when the best time would be for his family to come with an official proposal. During the day, they used text messages as a means to exchange their emotions and to up-date each other on what was happening in their respective lives.

> 'Teens use short messages to express the entire spectrum of human emotions. Through SMS, teens hate, love, gossip, mediate, and express longing, even when the writer lacks the courage to call or in situations where other communication channels are inappropriate. The text message is the back door of communication.' (Kasesniemi & Rautiainen 2002: 171)

In the case of Aisha and Samir SMS are vital in providing them with a way of communication that ca be more easily hidden from their families. Although his family generally does not mind Samir having a relationship to a girl, his parents would still try to reduce their communication mainly in order to save Aisha's (and also their own) reputation. Openly communicating with non-related, marriageable men is still not appreciated, often gossiped about and might result in severe difficulties for the girl's family to find a respectable husband for her. So, while most premarital couples might not even talk to each other when meeting in front of other people, they can still spend hours together in virtual spaces. In this respect, Reichenbach (2005) has illustrated well, how text messages offer a welcome opportunity for girls in Bahrain to establish relationships to boys and cross otherwise often strictly enforced gender lines. Using a pay-as-you-go service that offers free SMS upon topping up your phone, this is also an affordable way of communication between East and West London. Therefore, even after Samir's proposal has been accepted and their engagement has been celebrated, apart from the free phone calls at night, SMS remain a major means of communication.

For Samir's brother Salman computer-based communication technologies were far more important than text messages to communicate with Raya, even when both of them were in Mombasa. Fortunately, both of them had access to a computer with an Internet flat rate. While Salman could openly share the computer with his uncle, who was living on the same floor as him, Raya was able to use their computer at home pretending to do work for school. And the three months before the wedding that Salman was spending in the UK, in order to earn as much money as possible to contribute to his family's expenses on the celebration, they were completely left with this kind of virtual communication.

In recent years, researchers working on international migration as well as development researchers were struck by the important role mobile phones and the Internet have gained in facilitating and maintaining connections between migrants and their 'places of origin'. The use of International Communication Technologies (ICT) among 'transnational' migrants has since developed into a diversified subfield, having led for example to a special issue of Global Networks (2006, vol. 6, issue 2) in which the contributors address how and why 'transnationals' use particular communication technologies and how these effect network connections, communication patterns, senses of belonging and, following Escobar (1994: 214), even the relationship between the global north and south more generally (Panagakos & Horst 2006: 110). On the one hand, the diverse practices of ICT-use are seen to bring life to the more abstract idea of 'transnational space' by allowing for 'a sense of involvement in each others everyday lives' even across distances (Horst 2006: 156, see also Benitez 2006, Maintz 2008, Morley 2000, Rettie 2008, Tufte 2002). On the other hand, however, when taking a closer look, they also reveal the burdens, inequalities and obligations involved in transnational relations. In general, there is, for example, the obligation to call, the pressure to secure credit to be able to do so, the unavoidable sending home of money when being in reach of the steady demands expressed by relatives and friends. These, as well as remaining and maybe even enforced feelings of separateness, loneliness and longing, have however hardly been touched in the literature so far (for economic burdens see Mahler & Pessar 2001, Noguchi 2004, on emotions in transnational relationships see Walsh 2009).

When looking at Salman and his fiancée Raya, it is exactly these aspects of involvement, closeness and, at the same time, separatedness and distance that dominate their interactions prior to their wedding. At least when Salman stays with his family in their house in East Ham and Raya sits in their home office they can both still rely on an Internet flat rate and often talk for hours. Compared to many others who do not have a computer at home, this makes their communication at lot more relaxed and does not effect their personal finances as visiting an Internet café would do. Although in Salman's case an Internet café might sometimes even offer more privacy than using their home computer.

In the house of Samir and Salman's parents in East Ham somebody is always online. Mainly using the msn messenger, their parents regularly communicate with the grandfather and his wife in Mombasa. Apart from that, they also use the opportunity to exchange news with his father's sister and her family who moved to Toronto (Canada) in 2005, as well as with other relatives. This way it is possible to also discuss some more mundane aspects of each other's lives, as there is no clock counting the minutes one has to pay for. Often, it therefore happens that the messenger is on with an open conversation although nobody has typed a message for about half an hour, keeping each other in reach and performing a certain closeness or presence while still being able to do something else. Sometimes, however, this also occurs involuntarily, when too many people in the house try to chat with somebody at the same time, all insisting on continuing their conversation not willing to give way to the others for too long. To reduce the quarrels concerning the use of the computer, their father had bought a laptop, which is now generally reserved for him and his wife as well as for any of the children who has to do work or school related tasks. Whereas some flows of virtual communication only reach a couple of streets away (for example when one of the girls chats with a friend from school and with some of her cousins living in or near East Ham), others connect them to their friends and relatives in Mombasa as well as in other places such as Dubai, Dallas and Toronto.

As it is hardly possible to control this kind of communication – regarding the contents as well as the conversational partner(s) – instant messaging services serve as the perfect space for multiple, 'non-committed' flirting. Samir and Salman's younger siblings, both boys and girls, make extensive use of this, flattering, teasing, and wooing, often showing each other the responses and discussing what to say next. This situation does not always make it easy for Salman to communicate with Raya without his brothers and sisters wanting to join in the conversation or commenting on his expressions. Moreover, what bothers Salman is that his online communication with Raya is only possible when they are at home, but that there is hardly a way to inform each other of what they are doing while being away from the computer. This is especially hard during the days Salman spends in Wales. There, he buys telephone cards, using the landline of the place where he stays to call Raya in the evenings. In these conversations they can elaborate their dreams of their big wedding and indulge themselves in imaginations of their future life as a married couple.

Strategic talk: Material considerations on a family level

The two brothers Salman and Samir therefore both succeeded in their proposals, although at first having had to face an explicit resistance. While Raya and Aisha have always been sure about their choice and clearly expressed it as a decision of love, it was their parents and close relatives to whom economic aspects played an enormous role in their considerations of the respective proposals. Although, in both cases, the families eventually gave in and accepted their daughter's wishes, it becomes clear that, if it would have been at them to choose, they would have chosen their son-in-law according to different criteria. And, even though marriages that can look back at a preceding (secret) relationship are definitely on the rise, parents are indeed often still included in the search for a marriage partner. I regularly came across young men in East Africa, Dubai, and London who explicitly appointed their parents to help them finding a wife. In this respect, it can be observed that, whereas for relatives on the girl's side economic considerations often weigh heavily (but still cannot make up for an unpopular family background), to the parents of the son these usually only play a minor role in choosing a girl – though higher education and a job do matter, but more as an indicator of her character than as a financial asset.

The *harusi*, the wedding, is generally considered to be the main transformation of a child into an adult and thus is also the decisive step to become a full member of the society, while perpetuating descent groups as well as emphasising alliances between them (cf. Middleton 2004: 94). According to Middleton, whose main research on Swahili marriages dates back to the 1980s, no single wedding can be understood in isolation, but only as part of long-term strategies of families and subclans for commercial enrichment, increase in rank, the making of moral purity, and claims to social position (Middleton 2004: 95). In order to fulfil these aims, marriages of 'patricians' would generally be arranged between close kin (between paternal parallel cousins from the same lineage and between cross cousins from different but related lineages) so that both parties could retain their wealth, property and commercial rights within social proximity. As observed by Le Cour Grandmaison also in the late 1980s among the Omani nobility living in Zanzibar, this notion of social proximity has been extended to a few families 'known to be closely related', so that marriages were then ideally formed with partners not from the same kin group but of equal social status, economic position and prestige (cf. Le Cour Grandmaison 1989: 181). In both ways, marriages however have been regarded as long-term strategies, carefully planned and celebrated strictly in accordance with the social position of the families involved.

Until today, marriages among cousins or between close families are still very popular. But not only because the children deliberately try to follow their parents' ideals. On the one hand, this is facilitated by the relatively small Swahili communities that are dominated by large families and lineages, so that almost everyone is somehow related to everyone. On the other hand, it is simply more likely to meet and get to know each other when belonging to the same extended family. Among the weddings I have attended throughout the period of my research – at least fifteen in Pemba, Zanzibar, Dar es Salaam, Mombasa, Dubai and London – about half of the couples were cousins, and in the other cases, most of the families involved could at least build on already existing relations. Nevertheless, organising a wedding has definitely become more complex, according to the fact that many of today's couples and their respective families do not live in one place but are instead spread over different countries and continents. Although translocal marriages between Oman and East Africa have taken place for a long time, the majority of couples in the 1980s still belonged to families living in the same place, strengthening the relations within a community in a particular place. Moreover, while historical accounts particularly emphasise the strategic moment in arranging Swahili weddings, as I have illustrated above, even though there is indeed some strategic thinking behind the decision of whom to get married to – especially on the side of the senior women and men of the respective lineages – all the weddings I attended were based more on the decision of the couple than being the result of any betrothal arranged by the elders. In effect, what was much more apparent than long-term family strategies of how to retain wealth and status were the various strategies and ways of arranging a wedding following their children's choices, in which short-term logistics and present financial considerations play a much bigger role. The only thing that still seems to be entirely up to the parents is the negotiation of the dowry.

PUTTING SWAHILINESS IN 'THE BAG': MATERIAL EXCHANGE BEFORE THE WEDDING

According to my observations, the *mahari,* the dowry paid from the groom's side to the bride's family, varies enormously, from – in some families – around 10 000 USD paid to marry a woman in Dubai, to 3000 USD paid in Zanzibar and Dar es

Salaam or, according to Middleton (2004: 97) the equivalent of 1000 USD paid to the bride's parents in Mombasa, to the payment of a symbolic sum to fulfil the religious demands in less wealthy contexts or where the family objects to the idea of 'selling' their daughter. Apart from the dowry, the family of the groom also packs a bag filled with new things, such as clothes, materials for garments, curtains and bed sheets, a *chetezo* (incense burner), *udi* (frankincense) and other things that the bride will need to take care of herself, her husband and their new home.

With the weddings of two of her sons being celebrated only two weeks apart from each other, Samir and Salman's mother can combine the shopping for the two bags, an activity that lies completely in the hands of women. Usually, she takes either her daughters or one of her nieces with her to help her choose. It takes her almost two months of shopping mainly in East Ham, Green Street, Ilford, Finsbury Park, and a number of shopping centres in East London to get the majority of the things she wants to put in the two bags. In general, it is interesting to see how, to a certain extent, 'the bag' brings together material objects from different places within the translocal Swahili space, indifferent to the actual place of residence of the bride and groom, but sensitive to what 'Swahiliness' is seen to entail. To get a proper Swahili *chetezo*, for example, she asks a relative who will soon travel to Mombasa for a short visit to bring her one. Luckily, her sister-in-law spontaneously books a flight to Dubai to see some relatives in Abu Dhabi and agrees to buy some high-quality *udi* as well as two of the long black veils (*buibui*) for her. In a more detailed sense, however, the bag still enables the sender to add a specific 'local touch' to the 'translocal flavour'. With regard to the ready-made clothes, brand names on the tags, for example, serve as indicators of the groom's UK-connection. Whereas, in respect to Aisha's bag, it seems to be especially important to include many things that were obviously sent from East Africa, and by doing so, trying to evoke a sense of Swahili tradition, regarding Raya's bag, there seems to be a stronger emphasis on 'the latest arrivals' in the UK. Overall, their future mother-in-law also uses the opportunity to buy very similar and sometimes even the same things for both bags. In Kiswahili, wearing the exact same design is called *kuvaa sare*. This is often deliberately used to express one's relatedness or close friendship and is commonly seen at weddings or other communal gatherings. In this respect, buying some things in pairs does not only ease her job, but will also enable the two brides to publicly express their new relation to their husbands' family, at least during the time when they will all be together in the same place.

THE CHOICE OF THE WEDDING LOCATION

Within such a translocal context, in which many couples have not been living in the same place before the wedding and their relatives are spread over a number of different places, it is not that easy to decide where to get married. Celebrating the wedding ceremony in London or, for example, in Toronto usually excludes a big number of relatives from East Africa who would neither all be able to obtain a visa nor to cover the expenses for the journey. Especially for the older generation the effort is often too big, including a long journey and the (often) cold weather conditions in Europe or Northern America. Nevertheless, even when deciding to celebrate the wedding in East Africa only a minority of the people one is close to would actually be able to travel there. Because, for health reasons, Aisha's father would not be able to travel to East Africa this year, in the case of Samir and Aisha, they eventually decide to have the wedding ceremony in London. This way, many of Aisha's close relatives as well as Samir's immediate family will definitely be able to celebrate with them, as well as the members of the East London Swahili community. Against the background of having to arrange the wedding of Salman and Raya as well, they also decide to only organise a rather small wedding for Samir and Aisha in London. Afterwards, the family will then be able to travel to Mombasa, where a much larger wedding can then be organised for Salman and Raya, satisfying all the relatives who would not be able to make it to the first one. This will also be the one that relatives from the United States and Canada will be able to attend while spending their holidays in East Africa.

Especially during the summer months, when there are school holidays, probably at least half of the Swahili population living in London travel to East Africa spending their holiday together among their other relatives. In the case of Samir, his siblings and cousins, their journeys to East Africa are often supported and accompanied by their parents, who have a strong interest in their children (re)establishing bonds to Mombasa and also want to refresh their experience of the place themselves. But, although the ostensible reason for most of these trips is the visiting of relatives in order to revive and strengthen family ties to one's grandparents, aunts and uncles, for many unmarried young men and women, it is particularly appealing to get the opportunity to present themselves directly to relatives and friends at their age. For them, these holidays are generally a great time, when they are meeting or getting to know their cousins, often combined with little trips either into the interior or along the Kenyan coast. And, as the summer months are clearly the time when most Swahili living abroad find the opportunity to visit Mombasa, this is also the time when a lot of weddings take place, bringing together hundreds of people from different places, often serving like a marriage market to those not married yet. However, what is often hidden behind these frequent holiday trips to East Africa is the financial dilemma a lot of Swahili families have to face in order to make these journeys possible.

Also in the case of Samir and his family, most of his siblings and cousins do not have enough savings to pay for the flight and all the additional expenses. Therefore, as soon as the planning for the next trip begins, each of them starts looking around for financing opportunities, asking other relatives or even friends if they can lend them some money. One of Samir's cousins, for example, uses her student loan to pay for her ticket. Samir himself borrows some money from the husband of one of his aunts, who decided not to travel to East Africa that year, and Salman increases his working hours to be capable not only of paying for himself but also to support his mother. Some of their cousins who were originally also planning to travel with them eventually have to stay at home, because they simply do not see a way to afford the journey. Overall, the majority of the family members actually arriving in Mombasa at the end of July 2007, is either indebted, owing some money to a variety of people, or will at least be facing a considerable overdraft when coming back to London.

This however does not prevent any of them from appearing in new clothes and with presents, at least for their favourite relatives. Already weeks before the trip, everybody had started to get all the things together, buying new outfits according to the latest fashion, which they are keen to wear in Mombasa. The space left in their suitcases – often a considerable part of it – is filled with presents for friends and relatives in Mombasa. Mainly clothes, including lingerie, but also small electronic devices, such as a number of mobile phones, are squashed into their luggage. Coming from London, there seems to be no way out of performing to be a well-off 'translocal', which is most effectively done by wearing and bringing objects that travelled with them from the UK.

WEDDING CELEBRATIONS IN LONDON AND MOMBASA

Wedding I: Samir & Aisha

As soon as the decision was made that Samir would marry Aisha in London, specific locations were chosen for the different celebrations. To keep the costs at a minimum, their families decided that it was only for the wedding ceremony itself and the following reception that a hall was needed, whereas the *henna* party on the evening before the wedding, as well as the lunch on the day after the wedding, could take place in the house of Samir's family in East Ham. This meant that, due to the small space in this terraced house, compared to other Swahili weddings in London, they would only be able to invite a small part of the Swahili community to these two functions. Samir's family had already shown that it was possible for them to restrict the invitations to a minimum, when celebrating the wedding of his cousin in the relatively small banquet hall of an Indian restaurant in East Ham High Street. In the same hall, they also celebrated Samir's father's 50th birthday party. For Samir and Aisha's wedding they were however looking for a bigger place, so that at least for the main celebration they would be able to fit in a lot of people. They finally decided to hire the hall and the associated kitchen of the Jack Cornwell Community Association, located in Dersingham Avenue in Little Ilford. Suited for at least 120 guests and available at a reasonable price, this hall is also in easy reach from their place, extremely facilitating the preparations including the building of a stage for the bride and groom, the arrangement of the tables and chairs, and the delivery of the food. When looking more closely at the specific arrangements and various celebrations that form the core of this wedding in London, it will become clear how – through the careful use of particular material objects and communication technologies – it is intensely connected to the wider translocal Swahili space.

Henna Party (at home, Friday, 20.07.2007)

In the morning of the day before the wedding, Samir's mother, some young women of the family and I take the bus to Green Street, where we first go into an Asian salon. Here, they do 'threading' the way commonly practiced in Mombasa. Afterwards, some of them go to guy some shoes matching the colour of their dresses for tomorrow. Green Street is one of the most popular shopping streets of Swahili in London. Famous for Asian fashion and accessories, it offers many of the things also being *en vogue* among the Swahili. At the *nikkah*, the central wedding ceremony, for example, it has become very popular for Swahili women to wear a *salwar kameez*, consisting of trousers and a long blouse reaching down to the knees, usually called *Punjabi* among Swahili women. The taste in hair bands, fashion jewellery and shoes is also often very similar to the Asian customers shopping in Green Street. Just for the henna, that Swahili women get drawn for important occasions, such as weddings or the end of Ramadhan, neither the reddish henna nor the Asian design is much appreciated. To get a more Arab-oriented design they stick to members of the Swahili community to do the drawing and use *piko*, a black colour, to draw the contours of flowers and other ornaments that are then usually carefully filled with red henna. While in the afternoon some of the women go to one of their aunts who, besides henna painting, also offers waxing, the rest of us sits in the living room of the house of Samir's family, waiting for a distantly related Swahili woman to draw henna paintings on our hands, arms and feet, one after another. We just manage to get dressed in time to await the women from Aisha's family to celebrate the *henna* party with us. Aisha herself has been at her father's house the whole day, being prepared for her big day tomorrow. Although we will not see her before she will enter the hall, we are constantly in touch with her calling her or chatting with her to inform her about the proceedings.

Fig. 40: Getting prepared for the wedding

Often, big *henna* parties are organised by the bride's side, with 200-300 women celebrating, eating and dancing together to express their happiness about the marriage. The bride usually only attends the second half of the party, arriving late and then observing the celebration from her stage. Since this short before the wedding a bride is not supposed to be seen by anyone, on these big *henna* parties the bride is often completely veiled. Alternatively, smaller *henna* parties only comprising the closest female relatives and friends of the bride take place, in which the bride can more actively take part in the celebrations and might sometimes even join the dancing. One week before the wedding Aisha's closest female cousins had simply organised a sleepover. Furthermore, the women of both families had agreed to celebrate the night before the wedding at the place of the groom's parents.

When it starts around 9pm, women from Aisha's family arrive, most of them wearing a *dira*, a long wide dress of Somali design that has developed into a very popular dress to be worn at *henna* parties over the last years. The long *mtandio* (headscarf), sewn from the same material as the dress, can be tied around one's waist, emphasising the typical dancing moves of the women practiced at *henna* parties. Some of the women from Aisha's family also bring along drums to be able to perform some of the very old Swahili songs sung at this occasion. After a while of sitting together on the floor of the living room, exchanging thoughts on the arrangement and the organisation of the wedding, we therefore soon start singing and dancing. And even though most of the younger women and girls, who have spent a long time in the UK, are not very familiar with the lyrics of the songs pitched by the older women with the drums, they still know the melodies and enjoy the rhythm. Despite the small size of the *henna* party, which seems to be uncommon these days among well-known families, the women of the two families thus clearly succeed in celebrating the evening before the wedding by evoking a strong sense of a traditional *henna* party, dominated by the sounds of drums and the voices of elderly women. Later on, we turn on a CD with famous *taarab* songs named *zilizopendwa* (representing the Oldies and Evergreens), so that we can all sing along. At 11pm Samir finally arrives accompanied by his brothers, and while his brothers help us to hold him, we succeed in covering the groom with *henna* – another traditional practice at this event. This also marks the end of the party. Before going home, all the guests are handed over a little carton filled with Swahili dishes, some savoury snacks such as *samosa* and *bhajia* (fried potatoes), a piece of cake and a *halva* (a sweet confection), as well as grilled chicken legs that have been ordered from a *halal* fast-food restaurant nearby. This way even the food given to the guests does not differ much from the cartons that would be handed out at a wedding along the East African coast.

Nikkah and Kupamba
(Jack Cornwell Community Centre, Saturday, 21.07.2007)

Usually, the groom's side organises the *nikkah*, the main ceremony, while the bride's side is responsible for the wedding reception. Since, in the case of Samir and Aisha, the actual wedding ceremony will not be held in a mosque but also in the hall of the Jack Cornwell Community Centre, where the reception will take place afterwards, the two families join forces to manage the day together.

The time from 1pm until 4pm is reserved for the men who arrive in the hall after the *dhuhr* prayer. The husband of a niece of Samir's grandmother takes the role of the imam and beside him, the bride's father, the groom's father, Samir's brothers as well as some other very close male friends and relatives come together on the stage, where, as soon as everybody is ready, Aisha's father marries his daughter to Samir. Samir is wearing a white *kanzu* (long white dress) and a *kilemba* (Arab turban), and a dark overcoat according to Omani style. We – Samir's mother, some other women and I – are only able to see this ceremony by unobtrusively looking through the window in the kitchen door, separating us from the main hall. Still casually dressed, we are preparing the food in the kitchen. Around one hundred men have arrived by now, sitting on the tables watching the ceremony, and expecting food afterwards. While the guests still congratulate the groom as well as the two fathers to the now completed wedding, we start handing out big plates filled with *biryiani* (a famous Swahili rice dish, clearly informed by Indian cuisine) through a pass-through in the wall. Samir's brothers and some other young men then distribute the dishes on the tables, so that we do not have to come out and be seen by the men. Only when they have left, we quickly clean everything and rearrange the tables and chairs before we can go home and get dressed for the reception.

Fig. 41: Nikkah

Fig. 42: The newly-wed couple on stage

As announced on the invitation cards, from 5pm onwards the women arrive in the hall. Whereas at *henna* parties most of the dresses (*dira*) are sewn by tailors in East Africa or Dubai (if not by the women themselves), here, at the reception, usually called *kupamba* in Kiswahili, most of the guests wear long, tight, ready-made dresses bought in London. *Kupamba* literally means to ‚to adorn oneself' and this is the part of the wedding most closely informed by the style of US-American wedding receptions. Due to the exclusion of men, many women leave their hair uncovered, displaying colourful hair bands and big flowers. When the hall is full of women, the bride finally enters the hall in a long, white wedding dress. Slowly, Aisha is walking across the hall towards the stage, trembling a little as all the women have their eyes on her. Critically evaluating the appearance of the bride, her hairstyle, make-up, and dress, is a common feature of Swahili weddings, increasing the pressure that comes with being at the centre of everyone's attention. When arriving at the stage, Aisha first has to keep standing, waiting until the man with the camera has taken enough pictures of her in different positions. She then has to sit down, again letting numerous pictures being taken of her alone. Eventually, her closest relatives join her on the stage for a picture, and subsequently, the majority of the guests arrange themselves next to the bride to appear in a photo with her.

These wedding photographs and films travel far distances and will be watched in the living rooms of relatives in East Africa, Europe, Arabia and Northern America. Not being able to regularly see and meet all these people, this is a major way of making an impression – good or bad – on one's relatives and others. Knowing that is it not uncommon for mothers to suggest a possible prospective wife to their sons from having seen them in such a video or on a picture, the excitement especially of the young women about their outer appearance and behaviour seems quite comprehensible.

After a while of dancing and, by doing so, expressing our appreciation of the wedding, the arrival of the groom is announced. Now wearing a white suit, Samir enters the hall, accompanied by his parents, and followed by his brothers and sisters who walk in pairs. His brothers and a few cousins all wear black trousers and pink shirts, clearly succeeding in making the hoped-for impressive appearance. The way they self-confidently enter the stage, placing themselves behind the newly wed couple – Samir has meanwhile sat down next to Aisha on the sofa – they show their enjoyment about their performance. In general, the young men seem a lot more confident regarding their impression on girls, not caring as much about the opinions of parents as young women do. In line with gender differences in the Swahili context and as expressed in the comments accompanying the watching of wedding videos and photographs, this shows how the pressure to look beautiful has a much stronger effect on young (unmarried) women than on young men.

Now, with the joint couple on the stage, another round of taking pictures begins, starting with the next of kin and continuing until almost all the guest have been photographed with the couple. Whereas, for some women, the arrival of the groom and other men marks the end of the party, so that they soon start to leave, especially those guests closer to the couple remain there and start dancing together, either not caring much about their relatively revealing looks or even happy about this opportunity to appear dressed-up in front of their male relatives. Sounds of Swahili Hip Hop mixed with RnB fill the hall, and finally even the couple can leave the stage and join the dancing crowd.

Kombe la harusi (at home, Sunday, 22.07.2007)

The *kombe la harusi* is a lunch organised by the groom's family. To reduce the effort as well as the costs, Samir's parents have decided to put some white pavilions in their small garden behind the house. The next day after the wedding ceremony and reception, the women of the family thus spend the morning in the kitchen to cook Swahili dishes, providing the guests of this London wedding with the exact same food as they would be offered at a wedding in Mombasa. The *jelebi*, an Indian sweet made of deep-fried flour dipped in syrup, that is very popular among Swahili as well, have been ordered at an Asian bakery in Romford Road, where I pass by in the morning to pick them up. The men help to clean the garden and arrange the *busarti* (mats) on the ground. One pavilion is reserved for men, the other one is soon completely occupied by women. Only bride and groom sit on chairs under a roof in the middle of the two pavillions. Today, there is no music involved in the celebration. Instead, the guests talk about the celebration of yesterday, exchange the latest gossip and enjoy the food. When, in the late afternoon, all the guests have left, we start sending the first pictures of the wedding to Samir's grandfather and other relatives in Mombasa and Toronto.

Overall, going through the different celebrations involved shows how this wedding in London draws on a variety of different translocal connections and, by bringing them together, enhances a genuinely translocal atmosphere of home. The engagement with objects like Swahili cuisine and dresses especially illustrate the effort made to constantly (re)connect lives in London to lives in Mombasa. The performance of old Swahili songs at the *henna* party or the dancing to popular Swahili Hip Hop at the reception also emphasise the wish to evoke a certain sense of home which is further facilitated by the coming together of a big group of Swahili people and the use of the Swahili language. This sense of home is strengthened by making extensive use of the Asian presence in London in order to get things as closely related to 'Swahili culture' as possible. While this highlights the Asian heritage, importing certain things such as the *buibui* and frankinsence for 'the bag' from the United Arab Emirates also points at their Arab links. London clothes and American RnB further complement the picture, leading to a complex hybrid that is clearly recognised as a Swahili wedding by the guests.

However, these objects all have their material affordances and costs, so that the various practices and arrangements are always restricted by the financial means of the families involved in the wedding. In this case, costs are reduced by radically reducing of the number of guests and refraining from more ostentatious celebrations, particularly in respect to the reception, usually by far the biggest and most costly celebration. Since Samir and Aisha themselves, as well as most of their close relatives who are attending their wedding, have spanned their lives mainly between London and Mombasa, constant references are made to the way things would be done in Mombasa, how things would be cheaper there, and easier to organise. Evoking a sense of Mombasa, that – as we will see when looking at the wedding of Samir's brother – does not contradict but rather includes the other translocal references pointed out above, is therefore crucial in order to emphasise their translocal belonging that goes beyond London as their current place of residence and is constantly reattached to this place on the East African coast. In the following, by turning to the wedding of Salman and Raya, that starts only ten days later in Mombasa, I will now try to exemplify the role of translocal connections as they are expressed from a different perspective within the translocal Swahili space.

Wedding II: Salman & Raya

Instead of emphasising an atmosphere of being in Mombasa, as it has been the case in Samir and Aisha's wedding, the different celebrations that are part of Salman's wedding that is actually taking place in Mombasa, are rather characterised by a wish to convey a sense of their families' involvement beyond the East African coast. While in London, the attachment of the couple to Mombasa was what distinguished the wedding from other weddings in London, 'Mombasa' is now more or less given and nothing special, so what becomes crucial instead is to make those connections visble that epitomise the translocal nature of 'Swahiliness' and show how one is able to actively embrace these translocal connections as a family. By mainly concentrating on the *henna* party and the reception, I will show how, again, it is especially through material objects, such as decoration and dresses, that the translocal context of this wedding is expressed, and how cultural expectations thus always have to be negotiated with material affordances and the financial capabilities of the parties involved.

Henna Party (Baluchi Hall, Wednesday, 01.08.2007)

On one end of the hall, the family of the bride had built a big tent in gold and red with a cushioned seat in its middle and a golden cushioned heart on top of it. Next to the seat, there are two oversized golden jugs of Arab design, one on the left and one on the right, and, further left, outside the tent, stands an even more oversized incense burner (*chetezo*) painted in golden and green colour. That this design is supposed to evoke a sense of 'Arabian nights' is immediately clear to the guests.

Fig. 43: Staging an Arabian night (still pictures from wedding video)

Some of the guests arrive wearing their long black veils (*buibui*) and do not take them off during the whole celebration. This can have different reasons: It might be that they could not afford to get a new dress and would feel ashamed to wear old clothes; some of them claim religious reasons which would not allow them to be seen 'uncovered' on the video which might get in the hands of men; others express their more peripheral connection to the bride and groom by rather staying aside and wearing clothes that do not attract any attention. Whatever their motivation, those that keep on their *buibui* generally remain seated throughout the evening, observe the dancing party and leave right after the food has been served. Other guests, on the contrary, use the *henna* party to appear in far more revealing clothes. Some of the young girls in particular take the occasion to wear tight trousers, skirts and tops, a dress that would not at all be acceptable at any other of the wedding ceremonies and would never be worn in public. We, as the family of the groom, have decided to all wear a *dira*, but compared to the wedding in London, where this was the most common dress at the *henna* party, this time we are in the minority – hence, it makes us even more easily distinguishable as a group, the family of the groom.

At first, everyone takes a seat on the floor, curiously looking around and trying to figure out who else is already there. When the music changes from Swahili *taarab* to American RnB, some women start dancing. Our big group, consisting of the groom's mother, his sisters, two of his female cousins who travelled with us from London, his closest aunts and cousins living in Mombasa as well as some of their close cousins from Toronto, occupies a considerable space of the dancing floor. At a time when the atmosphere is already very relaxed, the bride enters the hall. In tune with the theme, she walks in to a traditional Arab song. Also according to the customs, Raya is veiled in a way that only her eyes are seen, as nobody is supposed to see more of her before the wedding, scheduled for the next day. After the usual photo shooting, the women start dancing again, praising her and the wedding through their active participation in the *rusha roho* (dancing while moving in circles). Then, the bride's best friend presents an Arabic dancing performance, putting herself at the centre of attention. Soon after that, RnB, Dancehall and fashionable Arab pop songs dominate again. The *henna* party is definitely the part of the wedding that allows for the most expressive dancing, and while most of the older women rather remain in the background, particularly the younger women do not miss this chance. Today, no sound of drums or traditional Swahili songs is to be heard.

That particularly those not currently living in Mombasa also enjoy the more 'traditional' aspects of a Swahili wedding becomes visible when, the next day, we proceed from the mosque to the hall where the groom will meet and take the hand of his bride (*kutoa mkono*). As the groom's family, we are part of the procession from the mosque *Masjid Mabruk* to the *Mombasa Women Hall* where Raya, her family, as well as the majority of the female guests are waiting for Salman. The sound of *siwa*, large side-blown horns of brass and ivory, as well as the loud 'ululuing' of the women accompany our procession to the entrance of the hall. It is evident that those of us not living in Mombasa particularly enjoy this procession and the public attention it creates. In London I had often heard them talk about the fact that in Mombasa roads are sometimes completely closed for weddings, involving the whole neighbourhood, while in London relatively small groups of Swahili celebrate marriages more or less invisible to the public. Even if we are not as many as to fill the whole road – the road in front of the Mombasa Women Hall is one of the biggest in the city – driving and walking through the city clearly recognisable as a wedding party, with the groom in front, provides an opportunity to enliven these nostalgic memories.

Kupamba (Diamond Jubelee Hall, Friday, 03.08.2007)

The day of the *kupamba* – also organised by the bride's side – all the women are indeed busy most of the day with *kujipamba*, to adorn themselves and get dressed up. Although the celebration starts after the *maghrib* prayer, we do not manage to get there before 9pm, as it takes a long time for all of us to get our hair styled, the henna painting renewed, and the make-up done. Therefore, when we arrive at the hall, most tables are already occupied. We nevertheless still find seats not too far away from the stage. The massive hall is filled with more than 300 women, some of them already dancing. The stage today is even bigger than it was on the previous days, with umbrella-like trees in pink and purple surrounding the white bench at its centre. Later on, a well-known *taarab* singer performs a number of songs, thereby indicating the organisational as well as financial effort made by the bride's family to make this not only a big, but also a pompous wedding. In front oft he hall, under a big tent, a huge buffet has been set up, offering the most popular Swahili dishes to the guests. While eating, listening to the performance of the singer, and dancing, more than an hour passes before the bride arrives. Today will be the last time for her to occupy such a big stage as a bride. And the massive attention that is directed to her makes her shiver a little when she walks in. But before she can sit down on the bench, she has to pose among the trees until the cameraman seems satisfied.

Almost an hour later, the groom arrives at the hall. When walking in, eight couples, composed of his siblings and closest cousins from London and Toronto, precede him. When approaching the stage they all part to the right and left of the carpet, leaving a pathway for the groom from the entrance to the stage. This time, all men wear purple shirts with a silver vest on top, bought in a shop in Finsbury Park in London. Like Samir in London, Salman walks in with his parents, wearing a white suit with a black bow tie. When later all the young men appear on the stage together, it turns out that Salman's brothers have handed over the pink shirts they wore at Samir and Aisha's wedding in London to the their cousins living in Mombasa. According to their limited financial means they had agreed to only buy eight of the purple shirts and vests. But, wearing no more than two different colours, they still manage to make a big impression – their shirts even matching the colours of the stage. While some of the guests start to leave soon after the groom's arrival, we, as the family of the groom, again start dancing together, also persuading some of the bride's relatives to join in. When Salman and Raya leave the stage to be taken to their honeymoon sweet in the White Sands Hotel – paid for by Salman's parents – they also dance a couple dances with us, slowly starting to relax from their third day in the spotlight.

Fig. 44: Newly-wed couple with male relatives (still picture from wedding video) *Fig.45: Posing for the camera (still picture from wedding video)*

One week later, a *kombe* will be organised at the house of Salman's grandfather, marking the end of the wedding celebrations. The day after that, Raya will move in together with Salman in one of the rooms already prepared for them on the second floor of his grandfather's house. It is planned, that a month or two after the wedding Salman will travel to London trying to arrange everything for his wife to follow him. Meanwhile, Raya will return to live with her family. But now, the whole family can first of all enjoy their holidays in Mombasa.

For Samir and Salman's younger brother Saleh, for example, this means to spend as much time as possible with Laila, a distant cousin, who lives with her family in Canada. The last time he had met her was at the wedding of another cousin in London, which both of them attended a year ago. Whereas in London they had only been able to see each other at the actual wedding celebrations, where men were not even allowed into the hall until the very end of the reception, this time he gets to see her more often, and at his brother's wedding he was even able to dance with her for a bit. Already since they had seen each other in London, they have been communicating via instant messaging services once in a while, but now, when seeing each other again in Mombasa, their conversations become more serious. And, whereas some of these 'holiday-relationships' might fade out soon after the end of the holiday, Saleh and Laila will actually get married two years later in Toronto at the end of September 2009. This shows again, how the flow of ideas, material objects and people in the context of weddings decisively contributes to the construction and constant reconstruction of the translocal space.

Looking at the celebrations of Salman and Raya's wedding, taking place in Mombasa, shows how the emphasis on translocal connections shifts in respect to one's location within the translocal space. Despite the common belonging to a translocal space, weddings are arranged differently according to where they are organised, always making an effort to evoke a sense of the distant and absent places. This is done with the help of particular material objects, such as – in the case of Salman and Raya's wedding – the oversized incense burner on the stage or the colourful shirts and vests worn by the young men at the *kupamba* – and are thus closely related to the costs of the wedding. In effect, how weddings in this translocal Swahili context are actually celebrated can only be understood with regard to the dialectics of economic considerations and cultural expectations, making it impossible to get to grips with this cultural practice by only looking at one of these aspects.

MAKING HOME AFTER THE WEDDING

Apart from finding a partner and deciding where and how to arrange the wedding, deciding where to live after marriage is a third aspect challenged considerably by the translocal context in which the cultural institution of marriage is situated. In the following, I therefore also want to point out the economic rationalities behind choosing a place to live, as well as processes of material exchange linked to these decisions and practices of home making in the respective places. Examining the material considerations and ways of making home as a couple gives an idea why these kinds of translocal connections can never be stable and ordered, and are thus better understood through a more 'rhizomatic' understanding of translocal spaces. By concentrating on the movements of members of one family in real time, the following section also tries to vividly account for the disorder and chaos that is entailed in the often rather spontaneous movements within this translocal space.

Finding a place to live

Coming back to London from Mombasa after the celebration of Salman and Raya's wedding in September 2007, Aisha moves in with Samir. Not having their own place to live yet, this means that they share a room in the house of Samir's parents. His younger brother Saleh, who has been sharing the room with Samir before, moves into the former living room. But, soon after they have all settled in, their landlord decides to sell the house, making it necessary for all of them to move out. Depending on housing benefits, the family is not completely free in choosing a new place to live, but instead has to choose from a list of houses published online by the local authority. A month later, they are told to move to another terraced house in East Ham that is supposed to be big enough for the nine of them. Samir and Aisha again get one of the rooms upstairs, with the room next to them prepared for Salman and Raya. The youngest son gets the third room, and the two girls have to share the smallest room, offering hardly enough space for a loft bed and a wardrobe. What has originally been designed as the living room becomes the room of the other two sons, and their parents take the room right next to the entrance. Having been close to them even before the wedding, Aisha soon gets used to her new family and the life as a wife. Not living alone with her husband reduces her responsibilities, as she now only has to look after one room and is able to share the cooking with her mother-in-law and the two girls.

Having moved into the new house, everybody is looking forward to Raya's arrival in the UK. However, while the room is being decorated for her and a new dressing table arrives to complement the furniture, they receive the news that Raya's visa application has been denied. Only a couple of days later, when hearing that Raya is pregnant, Salman thus decides to move back to Mombasa to live with her in his grandfather's house. And he is not the only one moving back. Not long after his departure, also his aunt rather suddenly decides to give up her life in London and, taking her younger daughter with her, moves to Mombasa. Her son Omar with his wife and their daughter follow her, and together they all move into a house of the family in one of the older parts of the town.

Particularly with regard to Salman's aunt, who has just finished her education and was working as a nurse in Whipps Cross University Hospital, this decision comes as a surprise. Just a couple of months before, she had still argued that she and her family would definitely stay in London until all their children would have finished their education, as this was their primary reason for being abroad. Now, her younger daughter had just started secondary school and was nevertheless already taken to Mombasa. Apart from the educational reasons concerning their youngest daughter, this sudden departure also seems to be in contradiction to her recent achievements, such as successfully completing her studies and being directly offered a full-time job at the hospital. Just recently, she had discussed plans with us to continue her studies and become a midwife. Her dream had been to return to Mombasa in about ten years to open her own dispensary, where she could do everything according to UK standards. Now, she thinks that with her education she should be able to find a less stressful job in Mombasa that would allow her to lead a more comfortable life than she would be able to establish in London. Indeed, only a couple of weeks after her arrival in Mombasa, she takes a job as the head of nurses of a section in the main public hospital. Although at first, her daughter finds it difficult to get used to her new environment – she had never lived in East Africa before – she also soon settles in, being surrounded by relatives and going to a secondary school following English standards.

With regard to Omar, leaving London also means to finally give up his attempts at university. Studying for a computer science degree at City University, while also working at Tesco Supermarket, he hardly managed to concentrate on his studies, and without the necessary preparation and motivation, he has not done well so far. Nevertheless, using the opportunity to receive a good education had always been his main reason stated for staying in London. The second reason mentioned was to earn enough money to create assets, which would then facilitate life in Mombasa. This was also what Salman and one of his uncles had been trying when going to work in Wales. Trying to spend as little money as possible in London, his uncle had stayed in the UK for almost three years, working as many hours as possible, favourably night shifts. Having two wifes in Mombasa, he regularly sent them money, but after a while, when realising that earning 'quick money' was not as simple and quick as perceived a couple of years ago, and the calls of his wifes for him to come back became more insistent, he had left London at the end of 2006. In 2008, when Omar's wife has just given birth to their daughter and he himself has neither proceeded with his education nor found a better-paid job, they thus happily agree to his mother's decision to move to Mombasa. Staying in a house together, they would be able to reduce their expenses while at the same time increasing their living standard. To earn some money, Omar starts to work as a kind of trader, carrying and delivering small goods and money sent between relatives from Mombasa to London, so that during the first months after their move he still regularly appears in at the place of his parents in East London. After a while, when this job turns out to not pay as well as he had hoped, he starts looking for a different form of trade in Mombasa.

When Omar's older sister Hanan also decides to travel to Mombasa to stay with her mother, their father is suddenly left alone in the big flat in London that they have been renting for many years. For him, the decision to leave London is not that easy. Despite a strong desire to live in East Africa and being closer to his mother and other relatives, he does not want to give up everything he has established in London without having any idea of what he would be doing to earn a living in Mombasa. Among their relatives living in London, he and his wife were among the few who managed to live without any government support. For more than eight years, he has worked as a driver, delivering fresh vegetables to the shops at night, working six days a week from 1am to 11am. Now, he is not willing to give this up in order to depend on the wealth and social status of his family-in-law. On the other hand, he is tired of this job, often preventing him from taking part in communal activities as he usually goes to sleep around 5 or 6pm to get up at midnight. On the long run, it would also not make sense to keep this big flat, deprived of most of its inhabitants and furniture – it would simply become too expensive for him alone. But while he is still continuing as before, weighing his different options, Salman's mother, his sister-in-law, who has just moved into the new house with her family, also decides to move back to Mombasa, taking her two daughters and her youngest sun with her to live in some of the spare rooms in the house of Salman's grandfather, her father-in-law. She had just been told to increase her hours working in a nursery in Stratford without getting a pay-rise. But while also having to look after the house and the children this seems to much to her. She is hoping to find a similar job in Mombasa, where housemaids would help her with the other work. Her husband, however, who has just started a Master degree in Business Administration and has a job at one of the big electrical retailers, is reluctant to leave London with her that quickly. Instead, he decides to give up the house and, together with the two remaining sons, they eventually join his brother-in-law in his flat who is happy to get some company.

With all the others moving away in different directions, Samir and Aisha move into a flat in West London, inviting Aisha's father, who has lived alone since his daughter got married, to stay with them. As her father would not return to East Africa, moving to Mombasa and leaving him alone in the UK is not an option for them. When Samir finds a job at a cargo agent at Heathrow Airport and Aisha gives birth to a daughter, they are very content with their new situation, even though they would not have anticipated this series of moves. Within a timespan of only one year after the two weddings, the addresses of most of the family members have changed considerably. While some have more or less suddenly decided to move to Mombasa, others had to move within London, so that

Samir and Salman's uncle has meanwhile remained the only one to stay in his place, formerly known as one of the hubs of their extended family life.

This shows how movements cannot be regarded in isolation but are generally part of a more complex mobile setting, with one person's mobility not being understood independent of others. Particularly young children are usually bound to follow their parents, but also for adults it might be the movements of relatives that serve as a decisive trigger or at least as a support for their own movement, as in the case of Omar and his wife, when it is the spontaneous movement of his mother that convinces them of doing the same. These social aspects involved in the decisions where to live however are always intertwined with economical considerations as well as political and legal constraints, here most clearly expressed in the case of Zakki's wife Raya who is denied entrance to the UK. As the sudden decision of Salman's aunt to move to Mombasa shows, the evaluation of the different aspects is never stable but constantly changes according to the dynamic context that always offers a multiplicity of possibilities. In effect, living in a translocal space seems to be most adequately characterised by spontaneity, short-livedness and instability, making it impossible to predict the movements and choice of the location of the individuals.

Only a couple of months after her departure from London, Omar's older sister Hanan, for example, moves back to London to live with their father. Divorced and disappointed from her last marriage, she has no intention of moving back to Mombasa again any time soon, but instead has the aim to take up her studies in Early Child Education again. But soon after her return, their father eventually decides to follow his wife to Mombasa, leaving her in the flat together with her uncle and two of his sons. This is where, in October 2009, we all celebrate Samir and Aisha's daughter 1st birthday. Most of the family members have just returned from Toronto, where Saleh's wedding ceremony took place. Even his mother has taken the opportunity to spend a couple of days in London before continuing her journey to Mombasa. Although at that time, it is still planned that Laila, Saleh's wife would soon follow him to London, it will eventually be him who moves to Toronto, allowing his wife to continue her studies. At the same time, his father, together with the only one left of the seven children who, three years ago, had all lived together with their parents in East Ham, leave London to at least temporarily join the rest of the family in Mombasa. London, having long been the city where most of the family members lived, seemed to have turned into a minor place of residence. But also this situation is not at all fixed and stable: Until today, in the summer of 2011, many moves have taken place, Salman's parents, his youngest brother, as well as his aunt and uncle have all returned to London, some of the other family members seem to spend as much in London as they do along the East African coast – every now and then interrupted by holiday trips and visits to family members living elsewhere.

Practices of home making

So far, I have illustrated the movements of different members of one family trying to give a vivid insight both in the mobility itself and into the considerations behind changing the place of residence, especially as a married couple. In a second step, I now want to come back to more theoretical considerations about practices of place making and the ways in which these places are made meaningful by their respective inhabitants. In particular, as I have stated before, I will thereby concentrate on the role of material objects in turning the new place of residence into 'home'.

As Duncan and Lambert have emphasised, in translocal networks the dual meaning of 'home' as both a space where daily-life is lived and as a space associated with a notion of belonging, may not coincide (Duncan & Lambert 2004). This makes it all the more important to evoke a sense (of the place) of belonging in the actual place of residence. According to Tolia-Kelly, it is the 'prismatic qualities of material cultures [that] ensure that these cultures become nodes of connection in a network of people, places, and narration of past stories, history and tradition' (Tolia-Kelly 2004: 314, see also Dyck 2005: 242, Valentine 2001: 86). In contrast to the ephemerality and situativity of discourses and practices, material objects are of a more permanent nature and are therefore able to evoke a more continuous sense of 'home' through their presence in the place of residence. In this respect, material objects incorporate the 'bonds established by people – individually and collectively – with the places where they live and lived, through which they pass, about which they think [...]' (Pascual-de-Sens 2004: 350).

The terraced house in East Ham, which has been inhabited in different constellations by Salman, Samir, their parents and siblings for a period of eight years before they were forced to move out, exemplifies well how material arrangements and specific material objects serve as a means to provide this typical English house with a sense of what can be labelled 'Swahiliness'. Whereas the three rooms upstairs have served as sleeping rooms, the rooms downstairs have fulfilled many different functions through different ways of its inhabitation. On the one hand being a living room, on the other hand serving as 'the guys' room', it has accommodated sitting as well as sleeping facilities, representative and decorative objects presenting the family background, and some personal belongings of the young men such as clothes, computer games, and CDs. During the day, it served as a living room for the reception of guests, as a dining room for women, and as a playing room for the children. The different functions of the room reflect attempts to retain a Swahili lifestyle. Ideally, a Swahili house would have enough rooms to cater for each of these activities separately: one room for the reception of female guests, one for the reception of male guests, one room for the boys of the family to sleep, and some extra space for the children. Due to the English architecture of terraced houses, it is however hardly possible to have such a house in London. Hence, as one still needs to be prepared for all these different occasions, certain rooms have to be used in different ways at different times, to always have space for the way things are usually done according to 'Swahili culture'.

On a Sunday afternoon, for example, paper tablecloth is put on the floor, big pots and plates for everyone are placed in the middle. While, in the presence of guests, the men leave the room to sit down next door and eat from a big communal platter, on other days they all eat together. Although the next room contains a dining table and chairs, they prefer to sit on the floor in the first room and use the other one more as a store or an extension of the kitchen, or – in the case of male and female guests – as the male dining room. As Petridou has argued, food is perceived through a combination of senses, and therefore serves especially well to 'evoke the experience of home as a sensory totality' (Petridou 2001: 89). Nevertheless, while for me Swahili dishes often evoked images of my time in East Africa, I got the impression that for them this is just the food they eat, rather than a deliberate evocation of other times and places. This is the food they grew up with, the food they know how to cook, food they like. As Swahili cuisine is composed of an abundance of different ingredients and spices with influences of Arabia and India, many staples can be cooked in a variety of ways. In London, they adapt to London's infrastructure. Making use of Indian and Arabic shops, *halal* butchers, fast food outlets, as well as cheap offers from Tesco or Sainsbury enables them to cook food according to their cultural tastes. While eating, we watch the video of their last family holiday in Mombasa. These visual images of their connection to Mombasa are discussed and commented on throughout the meal and, not only then but also in other occasions, seem to serve in a much stronger way to more consciously bring, in this case, 'Mombasa' into their lives in London. Photos of family members in different places are placed on top of the fireplace, growing in numbers every time a new child is born. Apart from these visual images representing their translocal connections, it is central objects of Swahili material culture such as *mirasho* (perfume flacons) – beautifully designed and used to sprinkle perfume among the guests, especially at the celebration of the birthday of the prophet Samir (*maulidi*), but also at a wedding or after a festive lunch – that point to the different regions important for Swahili identity. An Arabic coffee pot shows their Arab orientation to the guests, and after the meal a *chetezo* (incense burner) made of clay is used to distribute the smell of *udi* (frankincense) in the house. These objects are all imported from the Kenyan coast or Arabia when either one of them or another relative carries these things on one of their journeys. Besides, these objects also point out that it is particularly the everyday encounters with memories stimulated through different senses – vision, smell, and taste – that seem to be crucial in order to create a sense of home (in this respect see also Walsh 2006: 140 for a discussion of evocations of home among British expatriates in Dubai).

As another example from the flat of their aunt and uncle in London illustrates, 'home' is also created through maintaining certain textures, materials, colours, and designs. One day, their aunt came home with new sofa covers in gold. Apart from matching the golden and yellowish colours of the curtains, as well as to a new carpet, they also go well with the golden, shining *mirasho* (perfume flacons) placed on the shelf as well as the frames with golden qur'anic writing hanging on the wall. Following Edensor (2002), the term 'home-making' pinpoints the ways in which we 'make ourselves at home' in the world according to social and aesthetic conventions about conviviality, domesticity, and furnishing and decorating space. As Edensor has pointed out,

> 'there are recognised codes of décor and aesthetic regulation which are passed across generations and between locals and fellow nationals: the colour codes of interior domestic spaces, the styles of furniture, the range of artefacts and ornaments, the modes of demarcating domestic territorial boundaries.' (Edensor 2002: 58)

Their aunt's choice of how to decorate the living room clearly matches Swahili preferences of that time (as stated before in respect to the consumption practices of Swahili women, fashion in this context changes more or less yearly). But timeliness aside, shiny and golden colours are generally very dominant in Swahili interior design. Instead of going to look for new covers in an English shop, she thus went to an Indian shop where she can assume the items to meet her taste. Concerning décor, Swahili in London especially benefit from the numerous 'Indian', Islamic, and 'Arabic' shops in London, where they can get framed Qur'an verses, religious books and other decorative items. Whereas Law has argued that 'the absence of familiar material culture, and its subtle evocations of home, is surely one of the most profound dislocations of transnational migrants' (Law 2001: 275), in this respect, it seems that, with London being a translocality full of goods from many places in the world, Swahili rather easily find ways to decorate their interiors according to their taste and cultural belonging. Instead of 'London objects' giving the household a British touch, 'swahilisising' domestic space and inhabiting London's branch of the translocal community allows them to maintain the strived-for Swahili atmosphere.

On the other hand, when moving to Mombasa, Salman and Samir's aunt sends a whole container full of their furniture from London to East Africa, immediately giving the house in Mombasa a European touch. Apart from the sofas, beds, shelves and wardrobe, she also takes the big flat-screen TV, the DVD player, a laptop and the play station with her. This means, that her husband is not only suddenly left alone in the big flat, but also remains with little more than the bed and a wardrobe in his room and a small table and two chairs in the kitchen. The living room is completely empty and is only slowly refilled with a patchwork of furniture when his brother-in-law and his sons move in. When his daughter Hanan returns from Mombasa, she first of all has to buy new furniture for her room. This shows that, even when they decide to take the majority of their material belongings with them to East Africa, this does not necessarily mean that they also take them back with them. Certain things, such as furniture are seen to be cheaper in London, others, such as electronic devices, are regarded to be of better quality and still not more expensive when bought in the UK. Moreover, it would simply be too expensive to always send a full container of things from one place to the other, especially if their movements remain relatively frequent. Finally, things from London also serve as indicators for previous stays in the UK, as guests immediately notice their origin when being invited to the house in Mombasa.

Apart from using specific material objects to evoke a sense of a 'translocal home', referring to both the place of residence, the place of previous residence as well as the place (or places) of belonging, what is also very important are practices of bonding with other Swahili people living in the same place of residence. When Saleh moves to Toronto, apart from being with his wife, he is also immediately included in her Swahili community. In this respect, he starts spending a lot of time with her male relatives and their friends, tightening his relationships to people he knew before, but also developing new ones. For women, it is particularly through communal events such as wedding ceremonies and other religious festivities that they develop new ties or strengthen the ones to family members and other acquaintances living in the new place of residence. Although, apart from few exceptions, it is not the exact same people Hanan, for example, would meet on a wedding in London or in Mombasa, it would still mostly be members of the same families, making it relatively easy to deepen her relationships to them. This illustrates how people's mobility leads to the strengthening and deepening of ties between different related families as well as within families, without necessarily reaching far beyond them. Moving within a certain translocal context generally further strengthens its constitutive connections by tightening them and making them thicker and stronger. Fights and fists however can also lead to ruptures and breaks, often replaced by the development of new ties that contribute to the strengthening of translocal connections in other parts of the translocal space. Irrespective of these ties being harmonic or conflictual, to all the people I was in touch with during my research, they formed a decisive part in creating a sense of home in different places of residence. But, apart from meeting familiar and intimate people, and being around certain familiar and intimate material objects, such as décor, food, and clothing, what is also important to note here is, that there is also an economic rationale behind the preference for remaining within a specific translocal context. Similar to the idea elaborated by Middleton (1992, 2004) concerning the choice of the marriage partner, keeping within their own and a number of other, usually related families secures the context in which they are well known for a certain rank and social position. Certain family names are associated with a certain social standing independent of the actual economic success of its members, which, as I have shown, can be crucial especially in terms of marriage proposals. Moreover, and sometimes even more important, it is only the people one is close to (if by actual or imaginative relation) that one can ask for financial help and that are generally more open towards material exchange. Particularly in the context of holidays or journeys to attend a family wedding it is a common practice to borrow money from relatives and close friends, so that having this kind of close connections in the place of residence is a decisive prerequisite allowing for a physically mobile translocal live. And when someone living nearby is travelling, the opportunity is often used to ask the travelling person to carry some goods, if as presents or for business, allowing people to be actively involved in material exchange even if not travelling themselves. Effectively, these social ties thus also have a strong material dimension that essentially contributes to practices of home making in a translocal context.

6ᵀᴴ HALT: SEARCHING HOME IN A TRANSLOCAL SPACE

As this chapter has shown, translocal cultural practices also have a considerable economic and material dimension, including clothing, food, household furnishing as well as flight tickets. This perspective forms a contrast to the first three chapters in this part of the book, in which a special emphasis has been put on the cultural dimension of trading activities, generally considered as economic. In this respect, I have argued against a solely rationalist reading of trade as an economic practice, by illustrating how trading practices do not always follow thorough calculations and precise planning. Focusing on its cultural dimensions a central aim has been to demonstrate how economic calculations are not always at the core of translocal Swahili trading practices, but that family relations and processes of identification often play an even more important role instead. Moreover, a closer interpretation of these trading practices has also shown how they are often strongly informed by spur-of-the-moment opportunities, and how trade has thus to be regarded as an internalised practice, relating to the common conviction that 'this is how earning a living is done' and that 'this is what they (should) do', even if the economic benefit only remains small or might even be a loss. Nevertheless, 'culturalising the economy' does not imply to condemn any economic rationale in translocal trading connections. Instead, as I have argued in this chapter, pursuing a cultural perspective on 'economic practices' also has to entail reflections on the economic dimensions of 'cultural practices'. Put differently, examinations of the relation between culture and economy, a field usually termed 'cultural economy', also have to account for economy in culture, and this has therefore been the focus of this fourth chapter.

By taking weddings as an illustrative example, I not only concentrated on economic considerations, rationality, material constraints as well as material gains within translocal cultural practices, but also shed light on what it means to live a translocal life and the effort it takes to construct 'home' – not least through materiality and processes of exchange. From finding a partner to deciding where and how to celebrate the marriage, and finally, to find a place to live after the wedding, living in a translocal context poses particular challenges. In presenting two different weddings, one celebrated in London, the other one held in Mombasa, I was able to point out the complexity and costs of arranging a wedding in such a highly mobile and widespread translocal context. Similar to the organisation of the wedding ceremony, looking at practices of home making after the wedding has shown very well, how a translocal sense of home is particularly conveyed through material objects. Expressions of belonging are therefore often accompanied by material flows, signalling the translocal reach of the organisers or inhabitants. In contrast to Casey (1993:23), who emphasises home making as a way of anchoring or orienting yourself in a particular, singular place, these weddings and subsequent ways of establishing home can rather be interpreted as ways of engaging even deeper in a translocal setting, that explicitly goes beyond a single place. These practices of living and performing translocality, that indicate relations that reach far beyond London in Samir and Aisha's wedding while similarly expressing connections that go beyond Mombasa in Salman and Raya's wedding, however, are often constrained by the financial means and material capacities of the families involved. Hence, the organisation and celebration of a wedding is characterised by complex negotiations about what is possible (economically) and what is necessary (according to the cultural ideal).

As illustrated above, marriages in this context are often closely related to physical mobility, either if one partner moves to the other, or if the couple considers moving as a considerable improvement on their future prospects. These often seemingly hasty decisions to move are informed by reflections on where would be 'the best place to live' in a certain situation, considering personal preferences, the location of other family members, political and legal contexts (including the possible reception of governmental support) as well as educational and occupational opportunities (wage, pension, tuition fees and the availability of loans). Convincing intentions to stay in a place for long often dissolve due to flexible parameters, and changing evaluations, so that eventually things might turn out differently. What could simply be interpreted as indecisiveness or inconsistency can also be understood as an effect of not wanting to give up or loose any place, neither the one where oneself or at least one's ancestors are born, where the family is seen to belong to, nor the place where one has once lived. This means that, although economic considerations do generally play a role in translocal cultural practices, they cannot account for them alone. With every change of the place of residence, people deepen their translocal connections and expand them further, which also leads to an increasing mobility in-between these moves to visit friends and relatives in other places, so that translocality becomes more or less self-energising.

Changing references and preferences are drivers of ongoing translocal mobility, resulting in ever more fluid and processual translocal connections. Constant flows of people, objects and ideas do not only prevent the translocal Swahili space from being fixed and stable, but also make up its heterogeneity and deprive it of any clear order by constantly offering different opportunities between which to choose. For some, these opportunities are clearer or more limited than for others, who seem to constantly search the translocal space for the 'best place to live'. Moving from one place to another, deepening and expanding their connections while at the same time cutting off others, the image of a translocal space develops into a tangle and enmeshment, a 'rhizomatic' form that, compared to the classical network image, comes much closer to the way translocality is actually lived.

ENMESHMENTS

MOBILITY AND THE GEOGRAPHIES OF SPACE: CREATING A TRANSLOCAL SPACE THROUGH TRADE

The central aim of this third, and final part of the book is to enmesh the different lines of thought that have been developed in the previous parts. First of all, this will entail weaving together the different routes presented and discussed in the four chapters of the second part of the book. I will do this by examining the main cross-cutting themes, the first of them being mobility, which also served as the starting point for this work. By linking the different practices, discourses and experiences of mobility that characterise the four different types of connection, I try to provide a more holistic view of the ways in which mobility contributes to the constitution and constant reconstitution of translocality and the emergence of a sense of translocal Swahili space. Furthermore, by focusing on trade, in each of the four chapters I have attempted to shed light on different aspects of the specific relationship between economy and culture. Now, looking at the four examples on a more abstract level, I will reflect on a broader understanding of the interconnectedness of economy and culture in the translocal Swahili space.

Other recurring themes in the representation of the four different connections that formed the second part of this book have been the role of location, the matter of distance and the negotiations of Swahili identity within the translocal connections under study. By drawing on empirical insights gained into these issues, in a second chapter, I will finally approach the often very abstract discussions on translocality and closely related ideas of transnationalism, relationality and cosmopolitanism, grounding the rather abstract models that guide theoretical discussions on translocality in the actual lives of actual people will help to better grasp and account for the complex and manifold mobilities of people, goods, ideas as well as their spatial implications, and thus, to refine and enrich the understanding of translocality as a concept.

TRANSLOCALITY AS A LIVED EXPERIENCE

The overall question that has guided the research underlying this book is how mobility influences perceptions, constructions and creations of space. The Swahili case has been chosen as a significant context in which to do such research, since mobility has long played and continues to play a crucial role in the construction of Swahiliness and the creation of relations between dispersed Swahili people. On an empirical level, the main focus has been to get an idea of how Swahili people living in different places are held together through the mobility of things, people, and ideas, and how this both leads to and expresses a certain sense of space that goes beyond space as a bound unity and instead incorporates dispersed areas and the connections between them.

The term 'translocality' is seen to comprise both the focus on mobility and the interest in its spatial implications, and thus forms the core of this book. In my understanding of the term, I draw largely on the idea introduced by Freitag and von Oppen (2010a) who regard translocality as a phenomenon and a perspective. As an object of enquiry, it contains the sum of phenomena, which result from a multiplicity of mobility and transfers. 'It designates the outcome of concrete movements of people, goods, ideas and symbols which span physical distances and cross boundaries, be they geographical, cultural or political' (Freitag & von Oppen 2010b: 5). In this respect, translocality is neither reduced to mobility nor to any fixed outcome, but it incorporates the tension and interplay between mobility and situatedness, movement and stability, transgression and cohesion. As a perspective, translocality indicates a relational and dynamic understanding of the world, by focusing on connections and mobility and thus looking at place and space from the perspective of movements. Foregrounding the processual, complex and heterogeneous character of translocal relations, it leads to a cloudy or 'rhizomatic' image of the phenomenon, which – in comparison to the term 'network' that is far more often used to describe and analyse these kinds of empirical settings – is seen to better account for what living in a translocal context actually means to those involved, without framing them in a rather static and schematic image of points and lines (cf. Freitag & von Oppen 2010b: 20).

Bringing these two dimensions of the term 'translocality' together has led to an empirical study that aims at understanding the constitution of translocality as a central phenomenon in the everyday lives of Swahili people by putting connections at the centre of the research, and thus at investigating the translocal relations themselves. By carrying out multi-sited and mobile ethnographic fieldwork, I have attempted to gain a deep insight into the practices, discourses and experiences constituting four different forms of translocal Swahili connections as presented and discussed in the second part of the book. These can be seen as diverse links and paths through different parts of the world, which cross each other, meet and mesh, and by doing so play an enormous role in the construction of particular places and spaces. By weaving together the empirical impressions from the different connections, I would now like to elaborate a bit more on the specific ways in which translocality as a lived experience informs the conception of space and place.

THE DIALECTICS OF TRANSGRESSION AND SITUATEDNESS

As the representation and discussion of the four different translocal connections has shown, translocality is characterised by different dimensions and variations of mobility, and the ways in which these kinds of mobility are linked to attachment and situatedness. In this part, I will concentrate on human physical mobility, whereas the mobility of objects will be discussed in the interpretation of the relationship between culture and economy, and virtual mobility will be part of the discussion on relational space. Although the practices, discourses and experiences of mobility encountered along the different connections vary considerably, in bringing them together it not only becomes clear how, despite these differences, mobility always forms a crucial part of the respective ways of being in the world, but also how the assemblage of different mobilities brings with it a certain situatedness, a sense of stability and attachment that reaches beyond individual translocal connections.

When Mahir and Ibrahim undertake a journey through the Tanzanian hinterland, combining trading activities with visiting their relatives in the places they pass through, mobility is first of all characterised by newness and excitement, especially for Mahir, for whom this is the first journey into the Tanzanian hinterland. He experiences new means of transport and gets a new idea of distance and speed. By exploring a part of the world where he has not been before, the journey is an adventure full of new experiences, unknown landscapes and places. For Ibrahim, being able to experience all of this a second time – though this time as the guide – this journey creates a feeling of pride. The experience of travelling with a clear destination ahead somehow seems to also give, if only for a short period of time, a clearer direction to their lives, which expresses itself in the euphoric mood underlying their discussions of future possibilities on the journey, as well as the stories told about it afterwards. The duration of the journey, in this case about two weeks, therefore forms an exception to the everyday life of Mahir and Ibrahim, which allows them to familiarise themselves with new places, with relatives living in these places as well as with mobile trading practices. Moreover, it is a way for the young traders to find their way and establish themselves in translocal Swahili connections, thus their mobility also attaches them to the translocal Swahili space, providing them with a specific, translocal sense of belonging.

In comparison, business trips to Dubai can be characterised by their necessity and repetitiveness rather than newness and adventure. They are usually much shorter and might even last just three days, while at the same time demanding far more financial capital. Often organised rather spontaneously, these trips might occur up to once a month and therefore form a much more casual experience of mobility. Sometimes the physical mobility of the trader even remains unnoticed by friends and relatives. On the other hand, these trips are the core of the businesses of more professional traders like Majid, Khadija and Massoud as they are a necessary element in keeping their trading activities running. While regular flights make it relatively easy for these traders to reach Dubai in about six hours from Dar es Salaam, the bigger challenge usually seems to be the transportation of the goods, which is still much slower, more unreliable and makes it necessary for the shipment schedule to always be included in their calculations. The harbours, stores and shops show how this mobility is materialised in specific places and is therefore always linked to situatedness. But as the case of Majid shows, even shops are not necessarily fixed and stable but are moved from one place to another, indicating instability and insecurity in this otherwise very stable mobile life. Being able to afford and practice translocal mobility on such a scale on a frequent basis, traders like Majid, Khadija and Massoud live the translocal ideal that Mahir and Ibrahim are imagining for themselves on their journey through the hinterland, and that Seif and Matar are dreaming of on their daily cruises through Dar es Salaam. Whereas in professional trading practices mobility and situatedness are clearly distinguishable though at the same time being interrelated, in the activities of 'wannabe traders', mobility and immobility seem to refer to different kinds of mobility. On a physical level, the young men are all very mobile and are constantly on the move from one place to the next, to the extent that mobility can be regarded as a steady characteristic of their everyday lives. On a metaphorical level, they nevertheless seem more or less stuck in the ideology of a translocal life, triggering a considerable movement of mainly young men from Pemba to Dar es Salaam. Instead of *kusafiri* (to travel), mobility here mainly means *kuranda* (to walk around, though not necessarily by foot, to be 'out and about'), often without a clear destination, but as part of the daily restlessness and effort to earn a living. Whilst in both the business trips to Dubai and the trade journeys through the Tanzanian hinterland, mobility is seen to enhance the social status of the travellers, bustling about in Dar es Salaam seems to mainly affect the horizontal dimension with few variations in a vertical direction. The dialectics between mobility and situatedness is also expressed very well in the common Swahili expression *nipo nipo tu* (meaning I'm just there or I'm just around), which besides epitomising the often seemingly deadlocked and peddling nature of the lives of these young men, still alludes to their constantly being on the move (*kuranda* and *kuhangaika*) and thus goes far beyond idleness in a particular place.

With regards to the economic dimension of translocal cultural practices, I presented two different kinds of mobility – holiday visits (related to the attendance and organisation of a wedding) and moving house – that are both crucial in understanding the ordinary life of Swahili families and the ways in which they manage to keep together and deepen connections despite physical distance, in this case between London and Mombasa. In this instance, mobility is experienced as an opportunity (when moving to London promises a higher and more stable income, or when a holiday spent in Mombasa provides opportunities for romance), or as a burden (when one has to leave school and friends behind if one's parents decide to move, or when one does not know which place to choose as the future place of residence after marriage). But irrespective of this either positive or negative experience, mobility is generally taken as 'the way it is': a basic constituent of the lives of translocal families. Looking at the practices of home-making in this context shows how mobility is unavoidably linked to processes of materialisation and stabilisation. Furthermore, when trying to get to grips with the translocal mobility of traders as well as ordinary people, it is striking that this mobility is always closely related to the emotional attachment to the different places involved. This plays a decisive role in the choice of a place for economic investment as well as in the decision of where to live. In all of these translocal connections, mobility is regarded as a way of *kutafuta maisha*, which can be translated as 'to look for a living' but also – even more existential – as 'to obtain a life' (*kutafuta* – to look for, to search, to obtain; *maisha* – life, lifetime). This can be understood in the context of discourses on Swahili identity, in which the mobility between Arabia and the East African coast in particular is seen as the core constituent that has led to the emergence of 'Swahiliness'.

When looking at the different translocal connections presented here, it becomes clear how today's mobility is indeed closely related to the history of Swahili mobility. Travelling along the old caravan routes of Swahili traders and reviving the routes of the dhow trade between Arabia and the East African coast constantly evokes a sense of Swahili history. And it is exactly those references to what is generally perceived as a glorious past that feeds into the ideology of a mobile life, noticeable in the lives of the young men from Pemba who 'look for a living' but also 'look for a[n ideal Swahili] life' in Dar es Salaam and beyond. From an emic perspective, 'to obtain a life' as a Swahili and to establish one's position within what is perceived as 'Swahili culture' thus genuinely incorporates mobility.

Exploring practices, discourses and experiences of mobility along these four different routes shows how they not only differ in terms of duration, distance and direction, but also in aspects such as the level of excitement, routine, spontaneity, and purpose. However, in the context of all of these routes, mobility remains a central experience that goes beyond the different experiences of actually being on the move, and culminates in its representation of a crucial moment in the construction of Swahili identity. It is through different kinds of mobilities on different scales and in different directions that those on the move bring Swahili living together in different places, and therefore contribute to a sense of togetherness.

For example, when travelling through the interior of Tanzania, Mahir and Ibrahim get to meet some of their relatives whom they have not seen before. On his trips to Dubai, Majid gets the chance to bond further with Rashid and his family, and Badi and Manju develop their connections to their relatives in the United Kingdom. And also in the fourth chapter, it is the regular movements of family members between Mombasa and London that unite these two places in such a way that for some of them, moving from one to the other is not even experienced as a big disruption anymore. Seif and Matar, having moved to Dar es Salaam from Pemba, also contribute to bringing these two places closer together and by cruising through the city they contribute considerably to the sense of togetherness among Swahili people living in Dar es Salaam. And even if Khadija and Massoud prefer the Gold Plaza Hotel to staying with their relatives in Dubai, by travelling there, they still revive the longstanding and, for processes of identification, very important links between the East African coast and the Arabian Peninsula. In all of these cases, mobility is crucial in order to connect Swahili people living in different places and enmesh them in a translocal Swahili space.

However, apart from bringing people and places together, moving from one place to another not only fosters a common presence in a particular space but also always entails leaving people and places behind. The case of Mzee Farouk in Mpanda, who is left alone in his shop after his wife is taken to Oman for medical treatment and his sons have gone to look for business opportunities in other places, is a good illustration of how mobility also creates ruptures. The case of Samir and Salman's uncle, who is the only one to remain in London when his wife and children decide to move to Mombasa, also serves as a vivid example of how mobility implies that family and friends are at least temporarily dispersed. It exemplifies how connections change over time and how mobility leads to inclusion and commonality whilst at the same time bringing seclusion and loneliness. Moreover, as the case of the whole family illustrates, mobility is not only an expression of wealth and affordability, but it is also a sign of restlessness, dissatisfaction and an as yet unfulfilled search for home and stability.

Thus on the one hand mobility is a common denominator in processes of identification, which enables the connection of Swahili people in different places and forms the basis of a feeling of common attachment and togetherness. On the other hand, mobility always leads to ruptures, seclusion and loneliness – the former not being possible without the latter. Nevertheless, Swahili people rarely question the experience of the latter themselves. Even if Samir and Salman's uncle does not like being left alone in London, he would never hold his family back. Mzee Farouk might suffer from remaining alone in Mpanda, and Nabila might miss her husband when he is away on business trips, but these emotions will generally be subordinate to the sense of belonging to a translocal Swahili space, which is evoked through this mobility. By transgressing places and bringing different places together, these various kinds of mobility all contribute to a feeling that they are somehow all in the same space, a sense of space beyond a bounded physical surface.

TRANSLOCAL SPACES

By transgressing places and tying different places together, translocal connections enmesh and become entangled. This tangle, the enmeshing of translocal connections, is what I will refer to as a translocal space. As Rajchman puts it 'we *move* in space' whilst at the same time 'fill[ing] it out' with our movements, 'such that the space and our movements in and through it become inseparable from one another' (Rajchman 2000: 131). Correspondingly, Swahili people travelling between different places not only move in space, but also create space, which is translocal in the sense that it consists of translocal connections instead of a bound entity. Translocal space is made of connected lines and the meshing of those lines, not of surfaces. Therefore it always transgresses the local and can never be globally all-encompassing (cf. Latour 1993: 117–119). Moreover, as translocal spaces consist of moving people, goods and ideas, they change their shape according to the flows and places of interconnection and cannot be grounded in any fixed way (Massey 2004b: 5). However, they cannot be completely unbound from particular places either, indicating the dialectics between translocal and physical space. In constructing a translocal space we might try to abandon Euclidean space and the 'container view' that so often seems to come with it, but in doing so we also contribute to the persistence of Euclidean space, as otherwise we would not have anything to distinguish ourselves from. Though fluidity and processuality are intrinsic to any translocal space, they still float through specific places that can be found on a map, with some tributaries opening into wider flows, while others might dry out or be interrupted.

A vivid example of a flow drying out would be Mpanda, the small city on the Tanzanian mainland through which we passed on our journey from Sumbawanga to Tabora. Here, the number of Swahili residents is continuously decreasing, also leading to the ebbing away of translocal flows. Interruptions of translocal flows can be seen in places in eastern Congo or Burundi, where mainly due to political reasons, many relations have been cut off or at least extremely reduced. On the other hand, flows to places on the Arabian Peninsula, in the United Kingdom and other parts of Europe, to Canada and the United States have intensified considerably over the last twenty years. These examples show how translocal space evolves from ever-changing translocal relations, rather than being a clear-cut, enduring entity in physical space. Depending on one's position and individual relations in a translocal space, certain parts of it might seem more cloudy and fuzzy than others.

When examining the relationship between space and place from this translocal perspective, it becomes clear that there can be no conceptual differences between the two. When place is understood as an effect of multiple and heterogeneous relations that come together and intermingle, space is also this (cf. Amin 2002, 2004, Massey 2004b: 7), the only difference being that a space is usually seen to comprise more than one place, as well as the translocal connections holding different places together. While Mpanda, Macontainer (the shopping area in Zanzibar where Khadija has her shop), Magomeni in Dar es Salaam or East Ham in London (both areas with a relatively high density of Swahili inhabitants), are all places that play an important role in the everyday lives of the people living there, what holds these people together are the translocal connections which not only give the places their specific character, but also bring them together physically and imaginatively by evoking a diffuse, but still spatial sense of belonging. In a context in which translocal connections are extremely relevant in processes of identification, space therefore becomes extremely important as a reference. According to Casey (although he referred to place), 'what matters most is the experience of being in that [space] and, more particularly, becoming part of the [space]' (Casey 1993: 33).

But how does one become part of a space that evolves from translocal connections instead of the drawing of boundaries around a particular surface? While it cannot be sufficient to simply be in a certain place, even though this place might be pervaded by translocal flows, the crucial aspect must be to become part of these translocal connections. And, as mobility is at the core of translocal connections, this means that in order to become part of a translocal space one has to become mobile in one way or another. In order to belong to a translocal space, the key is not to belong to any specific physical location or territory, but to share and participate in the mobile practices that constitute it.

CULTURE AND ECONOMY IN A TRANSLOCAL SPACE

The focus of this book has been an examination of these kinds of translocal Swahili connections by taking trade as the empirical access point. In following four different forms of translocal connections, from professional traders to 'wannabe traders' and ordinary people, I have attempted to highlight the respective role of different kinds of material exchange in holding people and places together. Being a major mobile practice that constitutes translocal connections, participating in these practices is crucial in order to belong to the translocal Swahili space that emerges from them. This interrelation of trade and a translocal sense of space already indicate that what would usually be considered as economic practice (trade) and cultural ideas (translocal Swahili identity) is closely intertwined. In the Swahili context, it does not seem to make sense to regard trade as either economic or cultural. Instead, it seems more appropriate to see 'economy' and 'culture' as terms that indicate different aspects of Swahili trading practices. While theoretical debates on the relationship between culture and economy in geography (raised with the 'cultural turn' and particularly present in the field of economic geography) mostly centre on the dispute between '(political) economic' and 'cultural' approaches within the discipline, arguing which perspective – including its methodological implications – is better suited to analyse our contemporary world, it soon becomes clear that from an empirical perspective trading practices are informed by politics and the economy whilst at the same time being unavoidably 'cultural'. As they are carried out in a particular way and underlie certain ideas and values, they cannot be uncoupled from the cultural context in which they take place. Still, trading practices are also 'economic', since trade is based on buying, distributing and selling goods, all processes often motivated by the prospect of earning money. Thus, instead of foregrounding one perspective or possibly even playing one perspective off against the other, it is particularly interesting to tease out how cultural and economic aspects of Swahili trading practices intertwine and become partners in the constitution of the translocal Swahili space. Thus, by combining the empirical insights of the four forms of material exchange, I now wish to engage with the question of how trade – as an economic as well as cultural practice – essentially contributes to evoking the sense of a translocal Swahili space.

When considering what might make trade special in comparison to other forms of translocal connections such as virtual communication or visits, what first comes to mind is the materiality involved in trading practices. While the latter two only last for the duration of the talk or visit, material exchange enables the presence of goods in different places for much longer than the actual act of trading. Goods can be stored, displayed and might even become a constant companion in everyday life. As material objects are placed in shops, in houses or in containers, as they are being worn, used or cooked, it is through trade that the translocal space materializes. Since the majority of goods can be manufactured in high quantities, the distribution of goods to different places enables people to decorate their homes as a material site. The possession of the same objects in different places at the same time and the display of the same or at least very similar material belongings, is a visible sign that different people and places belong together. The difficulty in identifying the location of a particular event by relying solely on its material appearance – as for example in the case of the *kupamba* celebrations of the two weddings presented above – is a clear expression of this material dimension of the translocal Swahili space.

Moreover, the kinds of goods, their shape or style, indicate a lot about their (associated) origin and thus relate to specific places. The kind of incense used in the Swahili context can be immediately linked to Arabia, while cloves are usually associated with Zanzibar, irrespective of the place in which they are used. Material objects act as cultural signifiers, so that when selling a wooden chair built and carved in a style typical of Mombasa, a trader sells more than a wooden chair, he or she also sells 'Mombasa'. Through the acquisition of goods referring to different places it therefore becomes possible to bring these places together in one place, which then evokes the whole space through which these goods might have travelled. Material objects thus transcend their location, physically, but also symbolically, as they always evoke a sense of other places – a sense of Swahili identity. Hence, through translocal trading practices it is not only the traders that are actually mobile that become part of the translocal space, but also the consumers using the traded goods or placing them in their homes.

The important role of trade in the metaphorical constructions of the translocal Swahili space becomes particularly clear when looking at consumer practices. As my empirical insights have shown, consumption is more than a momentary practice of purchase; it is embedded in historically derived cultural preferences and ideas of 'Swahiliness'. Consumers are thus active participants in the construction of translocal space. This counts for the shopping practices of the more professional traders in Dubai, who carefully choose the goods according to what they know to be the current taste of their potential clients. It is the goods they pick in anticipation of their customers' tastes that will finally come to stand for 'Dubai', and it is the goods taken on the trade journey from Zanzibar into the interior that will signify the coastal relations of the Swahili people living on the mainland. A vivid example of the active construction of translocal space (this time on the part of consumers in Zanzibar) is the way Nabila has her own *buibui* (long black veil) made in Zanzibar. It matches the latest Dubai fashion perfectly and thus clearly fosters her translocal image.

Apart from the material and the metaphorical dimension of trade, it is also the very practice of travelling involved in trade and material exchange more generally that plays a crucial role in the constitution of the translocal Swahili space. Due to the fact that traders travel both to find particular goods and to search for customers, trade thus creates mobility and mobility also creates trade by inducing material exchange between people who have moved and now live in different places. This mobility in the context of trade brings with it personal encounters and visits of Swahili people living in different places, enhancing a sense of common belonging by frequent meetings and regular virtual communication. The sheer act of seeing or at least being in touch with others included in the organisation of translocal trade, is important in strengthening the sense of togetherness and practicing the translocal space. As emphasised with regards to 'wannabe traders' and their attempts to make a living through peddling trade, as well as in the context of the mobile phone biography presented above, people find many different ways to get involved in trading practices – or at least to create an atmosphere of trade. This shows that trade has not only long been, but still is, a widely shared practice, which is of high identificatory value, particularly due to its crucial role in constructions of Swahili history. It thus constantly fuels the idea of 'the Swahili' as a mercantile and highly mobile community. Overall, the translocal Swahili space therefore becomes constructed through the trading practices themselves, through shared material objects as the result of translocal trade that both materializes and symbolizes the translocal space. It is also constructed through imaginations related to trading practices in the Swahili context that strongly emphasise a common Swahili identity.

However, this translocal space consists of more than just trading practices, it also becomes articulated and unfolds in other domains of everyday life, such as communication, visits and the manifold relations people build within this space. After these more general reflections on how translocal trading practices contribute to the emergence of a translocal Swahili space, as well as how translocal space induces material exchange, I would now like to return to the central question underlying this work that asks how this translocal space is actually lived and experienced, and how empirical insights from the Swahili context can contribute to theoretical conceptualisations of translocality.

LIVING TRANSLOCALITY: GROUNDING THEORETICAL CONCEPTS IN THE ACTUAL LIVES OF ACTUAL PEOPLE

As Freitag and von Oppen (2010b: 5) argue, most theorising about 'global mobility' in the widest sense happens 'from above' and often has not only a 'western' and elitist bias, but also misses much of the complexity and diversity of translocal interactions and connections. In my view, empirical research on translocality in the Swahili context offers a way to discuss concepts that have emerged in discourses on 'global mobility', by cross-reading them with the lives of ordinary people experiencing translocality. First of all, I wish to relate the experiences of living in a translocal space to the idea of 'transnational social spaces', already developed in the 1990s as a concept that better accounts for the effects of what has been termed 'transnational migration' (cf. Pries 1996, 1999a). Since transnational social spaces are often presented as rather homogeneous, I will scrutinise in particular the role of location in translocal space. Secondly, as translocality builds on a relational understanding of the world, I will also engage in the discussion on relational space, which has become one of the key debates in geography in recent years (cf. Amin 2004, Bathelt & Glueckler 2003, Jones 2009, Massey 2004a). Drawing on empirical examples, I will focus in particular on the matter of distance, an aspect that according to relational thinking seemingly loses its relevance and reveals a lot about the relationship between topographical and relational space. Finally, the concept of translocality has also been closely linked to ideas on cosmopolitanism, since cosmopolitanism is often thought to result from extended travel and translocal connections. In the Swahili context in particular, mercantile translocal connections have often been a major point of reference in denoting 'the Swahili' as a cosmopolitan people, both from emic and etic perspectives. However, alluding to contemporary negotiations of Swahili identity, I wish to pursue a different perspective, arguing that a look at actual practices reveals how living in a translocal space leads to more inward-looking perspectives instead of fostering cosmopolitanism.

FROM 'TRANSNATIONAL SOCIAL SPACES' TO TRANSLOCAL SPACE: REFLECTIONS ON INNER STRUCTURES AND THE ROLE OF LOCATION

In discussing our theoretical understanding of 'translocality' this book is clearly more oriented towards theoretical and methodological reflections on the relations between mobility, place and space, than to studies of international migration. In the first chapter, when developing the theoretical context of this research, I engaged with different uses of the term 'translocal', clearly distinguishing it from the term 'transnational', which has become a key idea in migration research since the early 1990s (cf. Glick Schiller et al. 1992, Basch et al. 1994). Nevertheless, it is particularly in the context of transnational studies that efforts have been made to look at the effects of migration on place and the emergence of transnational spaces. In the following, I therefore wish to use these efforts as inspiration for the initiation of a more nuanced discussion of what I have introduced as a 'translocal Swahili space'. Since I am aware that the field of research on transnationalism is much wider and more diverse than I will be able to discuss here, I will mainly draw on some of the more prominent works on transnational space and point out common aspects and foci. These will then serve as a starting point to reflect on the kind of space that is constructed in this literature, highlighting some of the more prominent shortcomings when it comes to relating the concept of 'transnationalism' and in particular the idea of 'transnational social spaces', to the empirical evidence from my own research. In doing so, I will especially tease out the differences between translocal and transnational approaches to space and give a more vivid impression of the inner structure of translocal spaces.

The emergence of 'transnational social spaces' and the persistent emphasis on 'the nation'

The idea of transnational space has been most prominently made by Pries, who developed the concept of 'transnational social spaces' from his empirical research on Mexican migration to the USA (Pries 1996, 1999a, 1999b, 2001a, 2001b, 2008). Concentrating on the migration of Mixtecos to New York, he was struck by the strong ties between the new place of residence and the place of origin, which consisted of remittances, economic investments, political engagements, communication and regular visits. This allowed migrants residing far away to participate in the everyday lives and the celebration of public holidays of those who remained in the place of origin (Pries 1996). Based on these observations, he joins the calls for a change of perspective which acknowledges that international migration is not necessarily a unidirectional and singular movement from one place to another, but is instead embedded in the steady exchange of people, objects and ideas (cf. Appadurai 1990, 1996, Basch et al. 1994, Glick Schiller et al. 1992, Hannerz 1996). In his opinion, these transnational social links have reached such a level of 'gravity' that they are leading to the constitution of social spaces that are de-localised, spatially diffuse and that are unbundled from the space of nation-states (cf. Pries 1996: 463–464, Pries 1999: 3–4). In the following, he defines these 'transnational social spaces' as social 'interlacing coherence networks' (referring to Elias' 'Verflechtungszusammenhängen'; Elias 1978: 146ff), which are pluri-local and spatially diffuse. They constitute a social space which 'serves as an important frame of reference for social positions and positionings and also determines everyday practices, biographical employment projects, and human identities, simultaneously pointing beyond the social context of national societies' (Pries 1999: 26, cf. Pries 1996: 467). Four dimensions of these spaces are introduced as possible foci of research: firstly, the legal context of migration in the respective nation-states and the bilateral regulations between them; secondly, the materialisation of these spaces in communication, transport and organisational infrastructures; thirdly, the positionings in the social structures of the respective nations; and fourthly, the construction of (hybrid) identities as they are influenced by two different cultural contexts (cf. Pries 1996: 467–469).

What is striking here is how all of these different foci are characterised by an outward orientation towards the different national contexts, asking how 'transnational social spaces' are positioned in regards to fixed nation-states. Often ignored are the inner processes and dynamics characterising the 'transnational' space itself. This perspective seems to be driven by an interest in the effects of 'transnational social spaces' on the potential for integration and integration policies, which asks if these 'transnational social spaces' will eventually dissolve and give way to processes of assimilation or if they will develop into stable spaces 'crosswise' to nation-states (ibid: 469, Pries 2001a: 65). Moreover, the conceptualisation of transnational social spaces clearly expresses the sociological background of Pries, putting the role of national societies and the effects of 'transnational social spaces' on nation-states at the centre of attention, but hardly theorising the character of space itself. Drawing on a massive body of quantitative research, his aim is to prove the existence of 'transnational social spaces' and ask for political consequences, rather than being intrigued by the kind of space that emerges and what it actually means to live in it (cf. Pries 2001a: 56).

Today, nearly fifteen years after Pries' first publication on this subject, Google scholar identifies 867 hits for 'transnational social space' and almost 4000 entries for 'transnational space' (05.12.2011), spanning various disciplines and empirical contexts all over the world. Despite such variety of disciplinary and empirical contexts, what these publications still have in common is a certain notion of the 'national', not just because it is inherent in any use of the term 'transnational' (cf. Willis et al. 2004), but also by way of a more explicit orientation towards national contexts, as visible in the work of Pries. Even in the works of those who explicitly attempted to avoid this emphasis on nation-states by distancing themselves from this terminology and instead using the term 'translocal', nations often remain a clear reference. While this might indeed be helpful when focusing on 'problem analysis and problem solving' in the context of international migration policies (Pries 2001a: 65), processes of national integration or participation in national politics, I would argue that this orientation towards the national has clear shortcomings. Firstly, it excludes a number of communities that live in transnational/translocal spaces that cannot be described in national terms. Secondly, by undertaking an analysis of such spaces against the background of nation-states as fixed entities, it often fails to grasp their genuinely relational character. These two points can be well-illustrated by referring to the empirical context of the translocal Swahili space.

Translocality: Accounting for more than the national

In the majority of research on transnational (social) spaces, transnational subjects are named in terms of their 'original' nationality, either the nationality they had prior to their migration or, in the context of the so-called second or third generation, the nationality their parents or grandparents had prior to their migration. What seems most important is the nation-state from which they or their forefathers migrated. Although ethnic or regional foci are generally mentioned in the introduction to the empirical part of any publication, the title mostly indicates the national contexts of the research, such as from Mexico to the USA. However the problem is not just what appears as a simplification in the title, but rather the fact that it is often national contexts or particular places within these national contexts that are chosen first when designing a research project, foregrounding international (or sometimes even multinational) relations over actual translocal connections. Even if the actual research is later limited to specific communities, it hardly succeeds in escaping from this national framing. As a result, publications refer to transnational (social) spaces consisting of Mexican migration to the USA (e.g. Pries 2001a), New Zealanders in London (Wiles 2008), Filipinos in France (Fresnoza-Flot 2009), South Koreans in New Zealand (Collins 2010), Chinese in Canada (Waters 2010), Indians in Australia (Voigt-Graf 2004), or Singaporeans in London (Ho 2011). I certainly agree with Wimmer and Glick Schiller when they state that transnationalism should be regarded as an epistemic change of perspective, since transnationalism is not a new phenomenon but has long been hidden from a view that was captured by methodological nationalism (Wimmer & Glick Schiller 2002: 302). However this kind of 'national a priori' also represents a 'methodological nationalism' that has far too rarely been problematised in recent literature, which critically discusses the way in which 'the nation-state is reduced to the self-sufficient, solid and well-integrated representation of the modern society – when it is thought of as *the natural and necessary organising principle of modernity*' (Chernilo 2007: 18, see also Beck & Beck-Gernsheim 2009, Chernilo 2010, Wimmer & Glick Schiller 2002). Despite increasing reflections on the naturalisation of nation-states and ways to overcome this tendency, it is through relying on national terminologies when choosing the empirical focus of research that many of the connections going beyond national boundaries remain out of sight or are again read in the light of nations.

Concerning the translocal Swahili space, sticking to national categories would have meant talking about Omani in Tanzania, Tanzanians in the United Arab Emirates, Kenyans in London and Britons in Canada, among others. In this respect, the families being studied would be parts of different nations, each with a shared heritage or 'culture', or at least with a sense of national identity. According to Pries and others quoted above, looking at the Samir and Salman's family in London, for example, would have meant understanding them as part of the Kenyan social space, which – as an effect of the ongoing relations and movements between Mombasa and London – would become a transnational social space. Nevertheless in contrast to Pries, who generally assumes a former reciprocal exclusiveness between the surface of nation states and social space (cf. Pries 2001a: 53), it is impossible to assign Swahili people only one 'original' nationality or territory. Or it at least does not make much sense – in fact, it misses the crucial point. Swahili people are not necessarily better integrated into the national society of Kenya than of Tanzania, Oman or the United Kingdom. Although some members of Samir and Salman's family might indeed refer to themselves as Kenyans when asked about their origin in London, this can rather be explained by the fact that people have heard of it as a country – and a country is usually the expected answer – and not because they feel more Kenyan than Swahili. The same holds true for Swahili people with Tanzanian, Omani or Emirate nationality. This shows how such a perspective would hardly bring all of the people who have become part of this book together in one study, as it obscures the view on translocal dynamics by focussing on the links between individuals from a particular national society and specific places outside of this national space. As Larsen claims, 'in Swahili culture it is not important to be from the same locality', but it is crucial to be enmeshed in the same translocal connections (Larsen 2004: 123).

> 'For instance, a person from Lamu might not have been considered a member of the Twelve Tribes in Mombasa, but his larger identity as a Swahili, sharing common values and a common culture, would not have been in question. The larger fundamental "sameness" would have remained intact.' (Topan 2006: 64)

Relationality: going beyond the idea of a clear place of origin

The second point I want to make in respect of the study of transnational (social) spaces is that it fails to grasp the genuinely relational character of these spaces, by undertaking the analysis against a background of nation-states as fixed entities; one nation set as the origin and another one forming the receiving context of 'transnational' migration. This clear distinction between a sending country and a receiving country is usually understood to refer to stable entities independent of movement and an ongoing mobility between the two. As Basch et al. (1994: 8) state, 'transnational ties are taken as evidence that migrants continue to be members of the state from which they originated' (cf. Glick Schiller et al. 1992, Smith & Guarnizo 1998, Appadurai 1996). Though Vertovec (1999) extends this dualism between the sending and receiving context by developing 'triadic geographies of belonging', which differentiate the place of residence, the myth of homeland and the imaginations of diasporic communities, he still does not go beyond the idea of a clear starting point and a point of arrival. The idea remains that people have a place of origin from which they leave and then develop ties across borders from the place to which they have migrated. Although the image of networks is often used to illustrate how migration is embedded in existing ties spanning different places in different nation-states, one place is still considered to be the home country. This is illustrated for example in Boyd's early statement on the network approach in migration. According to Boyd (1989: 641), 'networks connect migrants across time and space. Once begun, migration flows often become self-sustaining, reflecting the establishment of networks of information, assistance and obligations, which develop between migrants in the host society and friends and relatives in the sending area. These networks link populations in origin and receiving countries and ensure that movements are not necessarily limited in time, unidirectional or permanent.'

Almost twenty years later, this image of transnational networks is still dominant. Though 'transnational networks [are seen] to have become more dense and efficient, [they are still regarded as] linking the sending and receiving societies' (Hollifield 2008: 78, see also Castles 2008: 54), with 'some being more oriented toward sending countries while others more toward receiving countries' (Zhou 2008: 237). Sending and receiving countries are still mostly conceptualised as fixed entities linked by network relations, which are then labelled as a transnational space. In contrast to this understanding, the situation of those Swahili presented in this book shows how it is the translocal connections that make and shape the different places involved in what then emerges as a translocal space. There are no fixed entities that later become connected by flows of people, objects and ideas, it is rather these flows that expand, interlace or become cut off, they constantly (re)constitute a translocal Swahili space in which a single place of origin cannot be discerned.

In Swahili studies, a heated debate on the origin of 'the Swahili' has been going on for many decades, with some scholars arguing that they are of Arab origin while others proclaim Africanness (Mazrui & Shariff 1994, see also Eastman 1971, de Vere Allen 1993, Middleton 2004, Topan 2006). It seems that the arguments are often more politically motivated than being an actual representation of the 'true origin' of Swahili people. While this academic debate tries to define Swahiliness from an outsider's position (though some of the scholars are Swahili themselves), focussing on perspectives from the inside frequently depends on the person one is speaking to, so that defining a place of origin often does not become any clearer. For example Ibrahim and Mahir's grandfather, who was born in Oman and moved to Pemba in the early 1940s, always considered his origin to be Oman. However, only some of his children and grandchildren follow suit, while others think of Zanzibar as their home. For members of Khamis and Rashid's family, scattered over different places on the Arabian Peninsula and the East African coast, it becomes even harder to name one single place of origin. The place of residence is the home to which one has returned, the place where one grew up remains the home one has left. And in the case of Samir and Salman's family, whose members continuously move between Mombasa, London, Toronto and Abu Dhabi, it becomes nearly impossible to differentiate between a sending and a receiving context, since various places are always both at the same time. What I wish to point out here is that a perspective that attempts to grasp connections and spaces beyond the boundaries of nation-states, needs to get away from the idea that there always exists a place of origin and a place of reception, usually in the form of nation-states. Instead of taking one particular nation-state as the home country, the Swahili context highlights how a translocal space is characterised by people's origins always becoming translocal too.

Homogeneity and Heterogeneity in translocal spaces

This focus on national categorisations and the adherence to one home and one receiving country often results in research that focuses on the relations between two places, which is then also reflected in the constitution of the transnational (social) spaces. In this kind of 'transnational' research, empirical research generally remains to be conducted in these two places, while the *trans* is often left open and unclear. It is imagined from information gained by surveys and interviews or from what can be seen in national statistics, but it is rarely explored in more detail and is generally reduced to the experience of living in a particular location, whether in the 'home' country or abroad. On the other hand, there exists another strand of research on transnational migration, which foregrounds the continuous circulation and transgression of boundaries in transnational space, presenting an 'image of transnational migrants as deterritorialised, free-floating people represented by the now popular adage "neither here nor there"' (Guarnizo & Smith 1998: 12). Ong and Nonini (1997:12) see this as part of an 'American cultural studies approach' that 'treats transnationalism as a set of abstracted, dematerialised cultural flows, giving scant attention either to the concrete, everyday changes in people's lives or to the structural configurations that accompany global capitalism'. In my opinion, both these perspectives – one emphasising homogeneity and harmony in translocal spaces, the other built on a more political economic view – obscure the everyday experiences of living in a translocal space. The empirical insights presented in this book highlight how translocal connections transgress boundaries and indeed lead to a common sense of translocal belonging that goes beyond national territories. But they also show that this neither means that there is a lack of heterogeneity and hierarchisation within this translocal space (cf. Willis et al. 2004: 10), nor that place and location no longer matters. However this heterogeneity and hierarchisation cannot be grasped adequately by a focus on national contexts and their location in either the Global North or the Global South, or by solely relying on political and economic factors more generally. On the contrary, examining the translocal Swahili space shows that these hierarchies and heterogeneities are often much more subtle and complex and do not stand in opposition to a certain kind of homogeneity.

In exploring the role of trading practices in the constitution of the translocal Swahili space, I pointed out how a translocal sense of belonging and togetherness is evoked through the exchange of material objects and the shared ideologies of trade and mobility. These aspects suggest a certain homogeneity within the translocal space. When sitting down for dinner in Tabora, the former trading post along the caravan routes into the African interior, we eat the same food, smell the same smells, wear the same clothes, and look at the same pictures on the wall as we do when having dinner in Dubai. Young Swahili men in Dar es Salaam and London share the same thoughts concerning marriage and are exposed to very similar expectations by their parents, girlfriends or wives. And when entering a hall in East London, where more than two hundred Swahili women celebrate a wedding, it is often hard to recognise if one is in Mombasa, Zanzibar, Toronto or London. But even though places might be exchangeable to a certain extent, where one actually is still makes a difference. Different places gain and lose meaning over time, as they are continuously changing depending on the translocal relations that pass through, merge and mingle. While the representation of the four different routes has exemplified how places along the routes participate in – and are constitutive of – the translocal space, it is also crucial to go beyond this perspective and look at how different places are positioned within this space. Of course, these hierarchisations and heterogeneities are not fixed but are subject to ongoing negotiations. They become visible in conversations accompanying the traders on the move, and they are also expressed in the choices of the traders. For example, in terms of the places through which to pass on a trade journey, whom to visit on a business trip or where to live after marriage. But although connections to certain places are more attractive than others, it is hard to develop a general hierarchy between the places involved, as preferences clearly vary at different stages in one's life, and are generally not as clear-cut and stable as one might assume. Even though they are clearly subject to economic and political aspects, the hierarchies within the translocal Swahili space cannot be reduced to this. This becomes particularly evident when looking at the imaginative geographies from 'the coast' and 'the interior' on the respective 'other'. Other forms of 'othering' and hierarchisation in the Swahili context can be found in the relations between 'Arabia' and 'Africa', and between 'Europe' (including the USA and Canada) and 'Africa', referring to particular attitudes and (life)styles that are assigned to Swahili living in these areas.

In respect to academic discourses on Swahili identity, it often appears as if the places along the 'Swahili coast' as well as Oman, as imagined origins of 'the Swahili' on a more abstract level, are assigned a special position based on the assumption that these places are particularly loaded with emotion and play a more stable and durable role in translocal Swahili space. While this might indeed be the case from the perspective of Swahili people living in Zanzibar where the actual and imagined relation to Oman is omnipresent, talking to Swahili in different places shows how the evaluation of the places that seem to lie at the core of Swahili identity also differ according to the place of residence. Instead of identifying a clear-cut hierarchical structure, multi-sited and mobile research among Swahili people indicates that all views vary according to one's momentary location, so that for example the difference between 'the coast' and 'the interior' might look much bigger from Zanzibar than from Dubai, and the difference between Zanzibar and Dar es Salaam might be minimal from the perspective of someone in Toronto, but might be extremely significant to someone in Pemba. Moreover, a closer look generally indicates views that are significantly more different than generalised imaginative geographies signify. When listening to reflections on where to open a shop, where to buy or where to live, it becomes clear how each place within the translocal Swahili space is seen to have particular advantages and disadvantages. It is the ever-changing evaluation of these and the very dynamic and small-scale internal differentiation that accounts for the high mobility in this translocal space.

The variability of internal hierarchisations – always closely related to one's location – also counts for processes of exclusion. Whereas along the Swahili coast, it is still the differentiation between 'the Swahili' and African mainlanders, taking up the longstanding dichotomy of *ustaarabu* and *ushenzi* (see p. 81), that continues to be a dominant topic, this is far less important to Swahili living for example in London, where it might be more important for Samir, Salman, their siblings and cousins to distinguish themselves from being Somali. Although a consciousness of the perceived differences between Omani and 'Zanzibari-Omani' might be extremely relevant in daily life in Oman, it is only marginal for the majority of Swahili living in Mombasa or Toronto. As the places involved in the translocal Swahili space in different parts of the world do not form a single entity clearly opposed to a neighbouring entity, but instead offer different fronts along which processes of inclusion and exclusion are negotiated, it is almost impossible to identify a common 'other'. Generally, more effort is thus put into connecting Swahili people in different places than into actively excluding others. This shows how internal differentiations, heterogeneity and hierarchisations do not stand in opposition to a more abstract sense of commonality, which both characterises the translocal Swahili space and holds it together.

In effect, the literature on transnational (social) spaces provides a number of starting points for the discussion of translocal space. Firstly, the orientation towards nation-states and national categories – an orientation that can still be found in publications on transnationalism and transnational spaces today – raises the question of whether this is still adequate for getting to grips with connections that go beyond nation-states. Though the persistence of 'the nation' within transnational studies is by no means a new critique, the Swahili example shows how this perspective not only sticks to the nation as a constant point of reference and explanation, but also how it narrows the view extremely, so that one might actually miss out many of the translocal dynamics. Secondly, transnational (social) spaces are in most cases seen to develop between one place of origin and one or more receiving contexts. This view of fixed entities linked by ties conceals the relational nature of these spaces, which are genuinely translocal in the sense that it becomes impossible to distinguish a clear centre or source. Nevertheless, this relational perspective should not result in any image of translocal space as completely homogeneous or only comprising free-floating, highly-mobile people, objects and ideas. By contrast, the transgression of boundaries is always closely related to the (re)creation of distinctions between the different places (cf. Freitag & von Oppen 2010b: 6). However these are subtle and fluid and cannot be grasped by solely relying on general economic and political characteristics of the nation-states to which they belong. Finally, looking at internal differentiations as well as processes of exclusion, shows how the spaces that emerge as an effect of various translocal connections are significantly more different, flexible, dynamic, and at the same time durable, than any conceptualisation of transnational (social) spaces would allow. Although the term 'space' is used to indicate a certain coherence, the concept of transnational (social) space mainly stays on a social/sociological level, without examining what effects translocal connections might have on the nature of space. The edited volume *Transnational spaces* (2004) by Jackson et al has already raised this issue, complaining that 'there has been too little attention to the transformation of space' induced by transnationalism (Jackson et al. 2004: xi). Aiming at expanding our understanding of these processes 'by exerting our collective geographical imagination', they however mainly concentrate on different kinds of 'transnationalism', i.e. on opening the field to 'transnational commodity cultures' and including empirical examples from different parts of the world. But, strengthening the geographical perspective on translocality also entails delving deeper into the specific ways in which translocal connections lead to the emergence of space, and how this can be related to geographical reflections on the constitution of space more generally. The discussion of translocal space, particularly against the background of current theories of relational space, therefore forms the central focus of the following section.

RELATIONAL SPACE: THE MATTER OF DISTANCE

The central assumption framing the literature on transnational social spaces is, as Pries has put it (1999: 3), that 'the relationship between geographic space and social space is presently being redefined'. After sociological thought having long been dominated by the idea that social spaces are represented in geographical spaces, Pries claims that against the background of increasing flows of international migration, one social space now often spans more than one geographical space, so that the 'the congruence of geographic and social spaces' becomes disrupted and diminishes greatly (Pries 1999: 3–5, 27). Though drawing on Blotevogel (1995) in order to give an overview of geographical approaches to space and how they are related to the social, 'geographical space' here remains to be used synonymously with 'national space', i.e. the space of nation-states. In German Pries also uses the term 'Flächenraum' (Pries 2008: 77) literally translated as ‚surface space', clearly expressing a topographic understanding of space. Moreover, as this 'geographic' or 'surface space' can be filled with or transcended by either one or more social spaces, an image of 'non-social' space as a 'container' can be clearly recognised in his elaborations (cf. Pries 2008: 132–133).

From a geographical perspective, this might seem rather reductionist and ignorant of the longstanding discussions on conceptualisations of space that have accompanied the discipline since its beginning and that are recently experiencing a renewed interest, particularly with regards to a growing interest in relational approaches to space (cf. Amin 2002, Harvey 1996, Jessop et al. 2008, Jones 2009, Massey 1994a, Murdoch 2006). However, even beyond the discipline of geography, a more general turn towards space and an investigation of its conceptualisations can be observed, which has for example led to the recent publication of two edited volumes, both multidisciplinary, on 'the spatial turn' (Döring & Thielmann 2008, Warf & Arias 2009). In respect to geography, the spatial turn appears to mainly consist of a relational or topological turn, which can be characterised by the general aim to leave behind topographical thinking and with it the 'Euclidean, deterministic, and one-dimensional treatments of space inherited from the "scientific" approaches of the 1960s and early 70s' (Graham & Healey 1999: 623).

A relational understanding of space has also been the theoretical perspective of my engagement with translocality. In the following section, I will therefore look at how relational thinking has been applied to understandings of space, in order to explore these (re)conceptualisations in respect to what I have labelled 'translocal Swahili space'. Can we really grasp translocality as a lived experience without any reference to topography? And how is this renunciation of topography reasoned? Giving a brief idea of the current interdisciplinary discussions on the spatial turn, I will point out the role this reconceptualisation of space has been assigned by geographers. By concentrating on the seminal works of advocates of the relational or topological turn, I will briefly point out the main characteristics attributed to topographic space and the ways in which space needs to be altered from a relational point of view. Introducing topology, I will particularly focus on the effects this has on the role of distance. By drawing on empirical examples presented in the course of this book, I will then bring these conceptual thoughts together with the ways in which distance is actually lived and experienced in the Swahili context. Arguing that distance does still matter, even in a genuinely relational space, I will finally attempt to rethink the implications of a relational turn on the concept of space, relating it to some of the scepticism voiced in recent literature.

The spatial turn: Perspectives from geography

Looking at the two volumes already mentioned on the spatial turn (Döring & Thielmann 2008, Warf & Arias 2009), it becomes clear that the spatial turn is perceived and applied very differently across the various disciplines, and that geographers – though often acknowledged as initiators – are not necessarily the key players. A major aim of *Spatial Turn* (Döring & Thielmann 2008), a book published by two authors from the field of German literature and media studies, is therefore to shed light on what appears as a 'space paradigm' in the humanities and social sciences. Whereas the first part compiles contributions from various disciplines, the second part consists solely of contributions from human geography, with the majority of authors being associated with the geography department in Jena, Germany. This is complemented by contributions from Ed Soja, Nigel Thrift and Mike Crang. Ed Soja, who seems to have been the first to use the term in a subheading in his book *Postmodern Geographies. The Reassertion of Space in Critical Social Theory* (1989), is the only author to have contributed to both volumes. Though in 1989 'the spatial turn' is meant as not much more than a hint at the rediscovery of the spatial consciousness in 20th-century-marxism to be identified in the work of Lefebvre in the late 1960s among others (Soja 1989: 39, see also Lefebvre 1968), it is not until 1996 on the back of Soja's book *Thirdspace* that the spatial turn is given a more paradigmatic tone:

> 'Contemporary critical studies have experienced a significant spatial turn. In what may be seen as one of the most intellectual and political developments in the late twentieth century, scholars have begun to interpret space and the spatiality of human life with the same critical insight and emphasis that has traditionally been given to time and history on the one hand, and to social relations and society on the other.' (Soja 1996, blurb)

For some, regarding the 'spatial turn' as the achievement of geographers seems to have led to a considerable change in the general attitude towards geography as an academic discipline. As Arias and Warf (2009: 1) put it,

> 'Human geography over the last two decades has undergone a profound conceptual and methodological renaissance that has transformed it into one of the most dynamic, innovative and influential of the social sciences. The discipline, which long suffered from a negative popular reputation as a trivial, purely empirical field with little analytical substance, has moved decisively from being an importer of ideas from other fields to an exporter, and geographers are increasingly being read by scholars in the humanities and other social sciences.'

While such a statement could be expected to flatter most geographers (and might indeed do so in most contexts), taking a look at the geographic contributions to 'the spatial turn' in the German context reveals a lot of scepticism and discontent towards the movement (cf. Hard 2008, Lossau & Lippuner 2004, Redepenning 2008). While Soja states that most of the spatial turn eventually occurred outside the discipline of geography (Soja 2009: 24), Hard goes further to argue that, in the course of the spatial turn, geography is constantly being reinvented outside of geography (cf. Hard 2008). As Redepenning notes, the spatial turn is celebrated as a new discovery in the humanities and social sciences without taking any geographical scholarship into account (Redepenning 2008: 318). This can be seen for example in Schroer's *Räume, Orte, Grenzen* (2006), in which the author tries to develop a 'sociology of space'. Tracing the way from an understanding of 'absolute space' to 'relational space', he exclusively refers to sociologists and never to geographers (cf. Schroer 2006: 174). In effect, the spatial turn in the humanities and social sciences is only rarely seen to be in tune with discussions on conceptualisations of space taking place within the discipline of geography. That the arguments run in different directions is also well-expressed in another statement in the introduction of *The Spatial Turn* (Warf & Arias 2009: 1): 'Geography matters, not for the simplistic and overly used reason that everything happens in space, but because *where* things happen is critical to knowing *how* and *why* they happen.' While this clearly reveals a focus on topography and the fact that space is considered as important for cultural and social analyses (cf. Böhme 2005, Weigel 2002), geographers at present seem to be much more occupied with the renunciation of topography and the basic question of how space can be conceptualised (cf. Soja 2008, Hard 2008, Werlen 2008).

Turning away from topography in geographical thought

In geography, debates on space have usually centred on the quest for its most adequate conceptualisation (cf. Hard 2008: 265, Redepenning 2008: 318), which would then enable geographers to clearly define their 'differentia specifica' (Hard 1993: 71). These discussions have long been characterised by a reliance on a 'containered' view of space, in which 'space acts as little more than objective, external containers within which human life is played out' (Graham & Healey 1999: 626). Based on Newton (1687) and Euclidean geometry, this 'absolute space' is seen to exist independently of any objects and any action. As Bathelt and Glueckler (2003: 123) pointed out in respect to the field of economic geography, this has resulted in a conceptualisation of 'space as a separate entity which can be described and theorised independently from economic action" (cf. Jones 2009: 489, Werlen 2008). This idea of space as existing irrespective of objects and actions meanwhile seems to have given way to the concept of 'relative space', which refers to the general relativity theory developed by Einstein (1960). As Jones (2009: 490) put it, taking up the idea of 'relative space' has led to a focus on the 'behaviour of objects and events that constitute and reconstitute the multifarious fields of existence'. Especially in the field of spatial science in the 1960s and 70s, this understanding of space brought location, relations between locations and distance to the forefront of geographical analysis, as well as bringing the use of statistical and numerical methods (cf. Bartels 1968, Harvey 1969, Wirth 1979). Space, particularly in the form of distance, is taken as a central determinant in a number of geographical theories, showing that this turn towards 'relative space' did not entail giving up the metric of Euclidean geometry, as it is seen to sufficiently describe the structure of the Earth's surface (cf. Blotevogel 1995: 734). In effect, space has usually remained to be conceptualised as a two- or three-dimensional metric regulatory framework of objects that can be localised on the earth's surface. Soja (1989: 37), more drastically, characterises geography's treatment of space in the middle of the last century 'as [treating] the domain of the dead, fixed, the undialectic, the immobile – a world of passivity and measurement rather than action and meaning'.

Both concepts, the 'containered' as well as a 'relative' concept of space, therefore rely on a topographic understanding of space, which regards space as a section of the earth's surface upon which activities take place (cf. Graham & Healey 1999: 624, see also Jones 2009). Moreover, formulations that are still common, such as 'society and space' indicate that space often continues to be imagined as distinguishable from society (cf. Blotevogel 1995: 734, Hard 2008: 268). It is especially the 'ontologisation' of space, in the sense of assuming space as real and given, that has become the centre of criticism among geographers in the last two decades. In Germany, Redepenning therefore regards the correction of the topographic/positivist conceptualisation of space as the central aim of the spatial turn within geography; for geography to assert itself as a progressive and critical academic discipline (Redepenning 2008: 319) and dismiss the 'negative popular reputation as a trivial, purely empirical field with little analytical substance' (Warf & Arias 2009:1).

A first step in this direction has been to understand space as a theoretical construct and not as the primary object of knowledge (cf. Bathelt & Glueckler 2003: 124, Hard 1993: 55, Werlen 2008: 370). Werlen, for example, has developed a research agenda that focuses on subjects instead of space and, by doing so, aims at investigating the everyday making of geography, i.e. the everyday constitution of spatial conditions ('Geographie-Machen', Werlen 1995, 1997). This nevertheless still rests on an understanding of space as a particular part of the Earth's surface and a strong focus on the relation between structure and agency, where space/structure can be transformed by and also transforms agency, but still remains separate of the social. Especially against the background of an increased interest in mobility, networks, and more particularly actor-networks and the relations they are made up of, a number of scholars postulate 'a shift in analytical emphasis from the reiteration of fixed surfaces to tracing points of connection and lines of flow' (Whatmore 1999: 31). For example in his article on the *Spatialities of Globalisation*, Amin (2002: 386), claims that 'those using the scalar logic continue to see places as sites of geographically proximate links or as territorial units', but that if we understand 'places as sites in networked or virtual [...] spaces of organisation: of placements of practices of varied geographical stretch [...], we cannot assume that local happenings or geographies are ontologically separable from those "out there"'. As Crouch (2010: 6) has formulated it in his recent article on *Thinking Landscape Relationally*, what scholars have to realise is that 'we live space, not merely in relation to it'. Consequently, as pointed out by Jones (2009) who traces the lineage of philosophical approaches to space in geography, relational approaches to space aim 'to replace topography and structure-agency dichotomies with a topological concept of space [...] as encountered, performed and fluid' (Jones 2009: 492). This quotation illustrates clearly how what began as the geographic turn against positivist understandings of space and the general image of space as a discrete, clear-cut entity on the earth's surface, has resulted in a turn away from topography and instead towards relational thinking and topology (cf. Amin 2004, Castree 2003).

The relational turn in geographic conceptualisations of space

Doreen Massey's *A Global Sense of Place*, first published 1991 in *Marxism Today* and gaining increasing popularity due to it being republished as a chapter in her book *Space, Place and Gender* (1994a), can surely be regarded as one of the milestones in the development of a relational conceptualisation of place and space. Though primary elaborations on the role of relational perspectives mainly concentrated on understandings of place (cf. Allen et al. 1998, Amin 2004, Smith 2001), 'recent years have witnessed a burgeoning of work on "thinking space relationally"' (Jones 2009: 487). For example a themed issue in the *Journal of Economic Geography* in 2003 (vol. 3), discussed the 'relational turn' in economic geography and its implication on concepts of space (see also Yeung 2005, Sunley 2008) 'as a means of overcoming the teleological and undersocialised nature of past approaches' (Boggs & Rantisi 2003: 110). 'Thinking space relationally' has also been the motto of an issue of *Geografiska Annaler* (2004, vol. 86, 1), where the focus is put on the implications of this conceptual position for political aspects. Here Massey introduces reflections on what 'relational space' might mean, not only as 'one of the theme-tunes of our times in geography' but also as a general commitment still needing further exploration to figure out what this conceptualisation entails in practice (Massey 2004a: 3). This commitment is pursued further in *Poststructuralist Geography: A Guide to Relational Space* (Murdoch 2006). Concentrating on a number of key post-structuralist thinkers, Murdoch has examined how they might help geographers to 're-conceptualise the spatial in ecological and relational terms' (Murdoch 2006: 2). His primary concern is how post-structuralism's interest in heterogeneous relations – on which I have also drawn when introducing the image of the rhizome (p. 23ff) – might impact on geographical theories of space, and at the same time allow human geographers to reach across the human-physical divide (Murdoch 2006: 3). This is also pointed out by Jones when he argues that:

> 'relational thinking is a paradigmatic departure from the concerns of absolute and relative space, because it dissolves the boundaries between objects and space, and rejects forms of spatial totality. Space does not exist as an entity in and of itself, over and above material objects and their spatiotemporal relations and extensions. In short, objects are space, space is objects, and moreover objects with all this being a perpetual becoming of heterogeneous networks and events that connect internal spatiotemporal relations.' (Jones 2009: 491)

I have tried to show how such a space might be imagined when reflecting upon the enmeshment of translocal connections as a translocal space. Though the significant features of 'relational space' listed by Murdoch (2006: 21–22) can all be found there, what I also argue is that such a space can neither be completely unbound from particular places nor from parts of the Earth's surface, i.e. that we cannot completely turn away from topography. Not only do we constantly refer to Euclidean space when trying to distinguish relational understandings from it (cf. Schlottmann 2008), we are also regularly drawn back to topography in our research: physical distance clearly still matters in the processes we observe and take part in, it matters to the research subjects as it matters to us. Nevertheless as I have illustrated above, relational approaches to space explicitly attempt to replace topographical notions of space with topological conceptions (cf. Latour 1988, Law 1999). In the following section, I therefore wish to explore the ways in which distance is accounted for in discourses of topological thinking, trying to rethink relationality by cross-reading its common conceptualisations with my empirical insights.

Distance in topological thinking, distance in everyday life

> '[The] science of nearness and rifts is called topology, while the science of stable and well-defined distances is called metrical geometry.' (Serres & Latour 1995: 60)

Topology can be traced back to Leibniz' *analysis situs*, in which space does not exist by itself but is regarded as a system of object relations, an *ordo coexistendi* (Leibniz 1966 [1904]: Math. V 178ff, Harris 2005: 123). Unlike topography, topology does not map discrete locations or particular objects (cf. Belcher et al. 2008: 499). Instead, topological space is non-metric and generally characterised by processes such as stretching, folding up and inverting itself (cf. Deleuze & Guattari 1976, Harris 2005: 124). Probably the most widely known metaphor for topological space is the crumpled-up handkerchief stuffed in a pocket, which is used by Serres in his conversation with Latour 'on a different theory of time' (Serres & Latour 1995: 60). When crumpled, the handkerchief brings together the near and the far, as two distant points are suddenly close together. This indicates that topological thinking assumes the 'subtle folding together of the distant and proximate, the virtual and the material, presence and absence' (Amin 2007: 103, Harris 2005: 114). Distance is not regarded as metric but, as Mol and Law (1994: 649) put it, has to do with 'the network elements and the way they hang together. Places with a similar set of elements and similar relations between them are close to one another and those with different elements of relations are far apart.'

In recent years, it is especially through the use of new communication technologies that topographic distance is seen to have lost its influence. That 'the faraway' can appear very near and form an important part of everyday life, has become increasingly common. In this respect, Licoppe (2004) has coined the term 'co-presence', referring to the '"connected" management of relationships, in which the (physically) absent party gains presence through the multiplication of mediated communication gestures on both sides, up to the point where copresent interactions and mediated distant exchanges seem woven into a single, seamless web' (Licoppe 2004: 135). While the role of topographical distance apparently diminishes, it seems to be the quantity and quality of relations that is seen to be decisive in defining what is near and far. But do relations really blur the boundaries between absence and presence in such a way that relational spaces can be adequately conceptualised without any meaning being ascribed to metric distance?

Focusing on the experiences of living a translocal Swahili space, numerous research encounters reveal how distance does play an important role. Even though the regular flow of goods and information creates a certain proximity between the young men living in London and their relatives in Dar es Salaam, they nevertheless remain approximately 7500km away. And in the course of this book, a number of examples have been addressed emphasising how distance actually matters: e.g. the duration of the shipment of goods or the fact that, despite regular communication, traders do not have any control over how the trade of the goods they send from London or Dubai is actually organised in Dar es Salaam and how much money they will eventually be paid back. Moreover, a trade journey like the one undertaken by Mahir and Ibrahim would simply not take place if it did not involve travelling through distant places. Distance is closely related here to difference, involving for example the passing through of different places and landscapes. The experience of physical distance is what makes the journey exciting and adventurous. In respect to trading activitities in particular, it has also become clear that if it would not make a difference for goods to be here or there, there would be no reason to take them from one place to another. However stating that distance indeed forms a crucial element in the organisation of trade does not imply that there exists a linear relationship between distance and the intensity of relation, as is usually assumed in 'spatial science' relying on relative space. In contrast to theories in which long distances result in the decay of interaction, the Swahili context shows that this is not necessarily the case. Through instant messaging services, text messages and phone conversations, people who live far away are indeed important parts of one's everyday life, providing virtual company throughout the day. However, this regular communication also always comes with the feeling that the other person is not actually present; the awareness of physical distance is a constant experience. Talking on the phone or seeing each other through the screen of the computer is at the end not the same as actually living together and being physically close.

These examples illustrate how – mediated through absence – distance remains an important characteristic in the lived experience of translocality, thus questioning the persuasiveness of a solely topological conceptualisation of space. The translocal Swahili space is fluid, expanding in one direction while fading in another, it might thicken and narrow at the same time, making it impossible to compartmentalise it on a fixed and discrete area on the earth's surface. It is only by following a relational perspective that it becomes possible to grasp such a space, as it is made and constantly remade through its multiple and heterogeneous connections. However, from an empirical perspective, the denial of the relevance of topography as suggested in geography's topological turn does not appear to be adequate.

Fetching topography back in conceptualisations of relational space

Relational thinking has had a remarkable influence on geography in recent years, including on conceptualisations of both place and space. In this book, I showed how a relational perspective can be applied methodologically in order to get to grips with the way translocality is actually lived and experienced, by providing a detailed ethnographic account of various forms of translocal Swahili connections. While a relational approach to space indeed allows for a more complex and fluid understanding of emerging translocal spaces, taking the lived experiences of people seriously also indicates its limits. In this respect, I agree with Jones who claims that 'mobility and fluidity should not be seen standing in opposition to territories and we should therefore not be forced to adopt a "network versus territories" scenario' (Jones 2009: 494). Though an *a priori* assumption of translocal connections mooring in discrete topographic entities would, as studies on transnational spaces show, certainly miss much of the nature of translocal space, relational space should not be conceived as being completely non-metric either (Amin 2002: 398). With topological space never loosing its topographic correspondent, topological and topographical conceptions of space need to be combined. This is also emphasised by Murdoch (2006: 98–99) in his explorations on 'space in a network topology' when stating that 'we should accept that space is generated from within sets of (networked) relations but we should also recognise that these relations must be situated within broader contexts of movement and flux' (cf. Jessop et al. 2008).

As Hard points out, every conceptualisation of space comes with a specific research agenda (Hard 1993: 70). A central agenda of topological concepts of space is its turning away from topography, but mainly for reasons that are not necessarily incorporated in topography itself, even though they have in fact long characterised geographical conceptualisations of topographical space in general. Fetching topography back thus does not have to entail the rollback of positivist or scientific approaches to space, since both are not coercively mutually constitutive. Space can be thought of as relational, it can be imagined as an enmeshment of entangled lines, and still, as the lived experience of such a space illustrates, it does not have to be thought of as being beyond the metric. It is not topography that is the problem, but its often far too naïve readings of space.

MOBILITY AND COSMOPOLITANISM: LIVING THE TRANSLOCAL SPACE

I began writing this book by embarking on two current discourses on mobility, a theoretical one visible in geography as well as in the social sciences more generally, and a much older but very persistent discourse on mobility expressed in Swahili studies. In both discourses mobility is not only seen as a practice but also as a way to engage with the world (cf. Adey 2010, Cresswell 2006, Horton & Middleton 2000, Middleton 2004). As Adey (2010: xvii) puts it, mobility is 'a lived relation; it is an orientation to oneself, to others and to the world.' Experiences in mobility are seen to ease one's way through the world, not only practically and physically, but also metaphorically in the sense of a mobile, flexible state of mind (Vertovec 2009: 70ff). In respect to Swahili studies, the longstanding physical mobility of people, ideas and goods that lies at the core of what is perceived as Swahiliness, is commonly regarded to have resulted in a 'cosmopolitan character' (cf. Bang 2008, Saleh 2002, Sheriff 2008, Walker 2008, Prestholdt 2009). In tourist advertisements, in novels, in academic publications, as well as in discourses on Swahili identity, Swahili people are deemed to be 'cosmopolitan'.

After having provided a detailed impression of the lived experience of many Swahili people, in which I particularly concentrated on the translocal connections within a translocal space, I now want to come back to the two discourses on mobility that were the starting points of this book and bring this journey to an end by rethinking the relationship between mobility and cosmopolitanism in the Swahili context. Against the background of actual practices and a more nuanced idea of the nature of translocal space, I believe it is time to challenge what now appears to be a rather idealised image of Swahiliness, and to instead draw a more complex picture that also considers the limits of transgression in translocal lives. Whereas translocality indeed develops from an outward orientation concerning the flows of people, objects and ideas between different places, the creation and maintenance of a translocal space relies at least as much on an inward orientation and self-relatedness. It is more an engagement in a translocalisation of one's own community and self than an actual engagement with 'the Other'. Regarding Swahili translocality as a way to constitute one's home emphasises not only how translocal connections transgress the local, but also how they at the same time localise this transgression. Hence, living in a translocal space and engaging with many different places does not necessarily mean that one ends up being at home in the world and manoeuvering expertly in different social and cultural contexts (cf. Hannerz 1996 [1990]).

Swahili people as cosmopolitan?

'Cosmopolitanism' is a dazzling term often used to describe people who move about in the world and have lived here and there, so that their biographies include stays in different places. Cosmopolitanism is thus also often related to translocality, with translocal connections serving as entries into different cultures (cf. Hannerz 1996: 104, 108). In this respect, the Swahili people whom I accompanied in their translocal lives could indeed be termed 'cosmopolitan', as they are constantly on the move and have contact with different people in different regional and national contexts. However, as Hannerz points out in his article on *Cosmopolitans and Locals in World Culture* (1996 [1990]), there is more to cosmopolitanism than just being on the move. Hannerz sees cosmopolitanism as a specific way to relate to the interconnected and diverse world, a state of mind or, with a more processual notion, 'a mode of managing meaning' (Hannerz 1996: 102). This perspective is particularly characterised by a 'willingness to engage with the Other' (Hannerz 1996: 103) instead of being preoccupied with oneself.

As Simpson and Kresse observe, the 'adjective 'cosmopolitan' is frequently used as an epithet in relation to the Indian Ocean, often in a somewhat lazy sense to suggest population movement and the sharing and inter-mingling of cultural values' (Simpson & Kresse 2008: 2). This does also apply to the Swahili context, in which cosmopolitanism often seems to be regarded as an inherent characteristic of Swahili culture and identity. In the literature on Swahili as well as in accounts of Swahili people themselves, Swahili identity is rather unchangingly characterised by their history as merchants and cultural brokers between the interior of Africa and places across the Indian Ocean, which is generally seen to have resulted in the cosmopolitan character of Swahili people as well as Swahili towns more generally (cf. Bissell 2007, Middleton 2004, Pearce 1920). Swahili cosmopolitanism is thus generally reasoned by their position as long-distance intermediaries, which is held to depict an outward orientation and the cultural competence to deal with difference. However, what is often missing in these accounts is the fact that this common identification with the idea of Swahili people as travellers and cultural brokers can also lead in the opposite direction, fostering a sense of sameness and internal coherence. I therefore wish to return to a novel by Abdulrazak Gurnah, the author already cited at the beginning of this book, as the following quote from *By the Sea* (2002) illustrates well how the dominant reference to the mercantile and cosmopolitan tradition of 'the Swahili' is often used by scholars as well as by Swahili people in order to define themselves and therefore also to distinguish themselves from 'the Other', instead of actually signalling openness.

> 'In the last months of the year, the winds blow steadily across the Indian Ocean towards the coast of Africa, where the currents obligingly provide a channel to harbour. Then in the early months of the new year, the winds turn around and blow in the opposite direction, ready to speed the traders home. It was all as if intended to be exactly thus, that the winds and currents would only reach the stretch of coast from southern Somalia to Sofala, at the northern end of what has become known as the Mozambique channel. [...] For centuries, intrepid traders and sailors, most of them barbarous and poor no doubt, made the annual journey to that stretch of coast on the eastern side of the continent, which had cusped so long ago to receive musim winds. They brought with them their goods and their God and their way of looking at the world, their stories and their songs and prayers, and just a glimpse of the learning which was the jewel of their endeavours. [...] After all that time, the people who lived on that coast hardly knew who they were, but knew enough to cling to what made them different from those they despised, among themselves as well as among the outlying progeny of the human race in the interior of the continent.' (Gurnah 2002:14–15)

As this quotation shows, mercantile culture and cosmopolitanism can also be denoted as a means to enable Swahili people to distinguish themselves from others, particularly 'Africans'. And the discourses characterising the relationship between 'coast' and 'interior' – elaborated in the context of the trade journey through the Tanzanian hinterland (see p. 81ff and 97ff) – also point out how 'cosmopolitanism' is a concept used rather to justify the boundaries between the two than to argue for the opening of such boundaries. This is also expressed in Saleh's article on *Tolerance: The principal foundation of the cosmopolitan society of Zanzibar* (Saleh 2002), in which on the one hand he praises the 'spirit of openness and tolerance', but on the other hand clearly sees cosmopolitanism expressed in the ability to integrate 'the other' and include it into one's self-image instead of engaging with different perspectives. Thus, the image of 'the cosmopolitan Swahili' seems to be closely related to a certain romanticism underlying scholarly as well as public discourses on Swahili culture and identity, whilst also fitting into political arguments that build on the 'difference' between Swahili people and their various neighbours. However ethnographic insights into the actual practices characterising everyday life, make it rather difficult to talk about a general 'cosmopolitan' stance characterising the translocal Swahili space.

The practices of translocal Swahili traders correspond to a certain degree to Hannerz' example of market-women travelling from Lagos to London with dried fish tied to their thighs hidden by wide gowns. According to Hannerz, this cannot be considered as cosmopolitan as this would need 'to entail a greater involvement with a plurality of contrasting cultures' (Hannerz 1996: 103). I have also shown in the Swahili case, how a large proportion of trading practices serves to distribute goods that are crucial for living one's own culture and fitting in with Swahili taste. An 'intellectual and aesthetic openness toward divergent cultural experiences' (Hannerz 1996: 103), or put differently, 'a search for contrast rather than uniformity' is not seen to play a decisive role in translocal Swahili connections. Moreover, another important aspect of cosmopolitanism emphasised by Hannerz is the changing relationship towards home. By following their ability to make their way into other cultures, cosmopolitans do not loose their competence with regards to their original culture, but 'can chose to disengage from it' (Hannerz 1996: 104). This means that to the cosmopolitan, 'home' looses its ability to be taken for granted: one's own culture is not natural or obvious anymore, but becomes one way of many (Hannerz 1996: 110). In the Swahili context, translocal practices point rather in the opposite direction. Material exchange is used to maintain and strengthen translocal connections in order to retain a nearness to what is regarded as Swahili culture, and not to develop a distance towards it. Through translocal connections Swahili people make and keep 'home' in different places and they even take 'home' along on their journeys. Rather than developing a cosmopolitan perspective towards the world and giving up the idea of home as something to be taken for granted, living translocality here seems to have the opposite effect: the taken-for-granted home is even expanded, so that home comes to embrace the translocal space.

Localising transgression: Translocal but at home

Particularly in the field of humanistic geography, senses of place and home have played an important role in researching the experience of being in the world. In this context, home has often been understood as a particular location, a place with special meaning attached to it. Tuan, for example, refers to home as feeling a sense of attachment and rootedness (Tuan 1991), and for Seamon (1979) home is also an intimate place of rest. These connotations of home as a place where one is rooted, attached and rests make it difficult to include a sense of mobility in the notion of 'home'. Transgression, as in the translocalisation of people and places, has in this context generally been seen as a disturbance, suggesting the loss of home. As long as place, the basis of a sense of home, is related to a break or pause in movement (Tuan 2001 [1977]: 6), the relation between mobility and place has to remain a complicated one (cf. Tuan 2001: 182).

Already in 1976, in his book called *Place and Placelessness*, Relph states that the frequent moving of American homeowners reduces the significance of home. In his view, mobility does not allow for the experience of being inside a place, i.e. 'to belong to it and identify with it' (Relph 1976: 49). Instead, people develop an 'inauthentic' attitude to place, which 'is essentially no sense of place, for it involves no awareness of the deep and symbolic significance of places and no appreciation of their identities' (Relph 1976: 82). The mass movement of people therefore encourages the spread of what Relph calls 'placelessness'. Almost twenty years later, Augé (1995) develops the idea of 'non-places', sites marked by their transience and the omnipresence of mobility. These are for example motorways, airports, department stores or internet cafes, all places of high mobility in which people co-exist but do not live together (Augé 1995: 110). However Augé's suggestion that rooted and stable places are increasingly replaced by non-places, does not have the same negative moral tone that is apparent in Relph's work.

Augé simply sees a challenge posed to the conceptualisations of place dominant among humanistic geographers by increasing mobility, which requires a more mobile way of thinking. As Cresswell (2004) has pointed out in his brief genealogy of place, while the perception of mobility's effects on place has first lead to the mourning of a loss of place, this has meanwhile turned into a more positive reading, in which the ability to lead mobile lives makes 'place' and 'home' appear redundant (Cresswell 2004: 48).

Instead of following either of these two extremes, a third way attempts to combine mobility and home in such a way that they do not have to be seen as mutually exclusive. Even Tuan has argued that lines or routes can be as meaningful as the end points (Tuan 1978: 14), showing that humanistic geographers have not generally been as adverse to the idea of mobility as is often presented. In Tuan's view, the sense of home can also be evoked in mobile contexts through daily routines and habits (Tuan 1974: 242). In this respect, it is not through the place itself but through particular practices that a feeling of familiarity and attachment arises, and this is exactly what characterises the translocal Swahili space. For example, when the young traders I accompanied reach Sumbawanga in the interior of Tanzania, a place to which they have never been before, they soon develop a familiarity with the place, encouraged by encountering well-known practices and attitudes in the homes of their relatives. Mobility being a central part of their identity further contributes to the immersion in these places as if they were their own homes. The people I travelled with generally follow established paths and patterns; they find and cultivate the same practices, norms and values wherever they go within this translocal space; and through the frequent mobility of objects, people and ideas they make places travel, so that to a certain extent, senses of place coincide in different places. Instead of being reduced to a single location, home becomes a set of relations. Despite frequent or even constant mobility, the translocal space does not loose the characteristics of home, but extends senses of attachment and belonging beyond particular sites. Instead of producing extroverted, cosmopolitan cultures, translocality thus translocalises the sense of home, always including social exclusion, as this translocal home cannot be home to everybody (cf. Freitag & von Oppen 2010b: 4). In effect, living translocality, though built on mobility, does not mean any opposition to a clear sense of home. On the contrary, translocal space offers a way to be constantly at home despite being mobile. As Casey (1993: 300) has put it, 'home is the place where, when you have to go there, they have to take you in'. Since the majority of Swahili people I met only go to these kinds of places, they actually never leave home.

Living translocality: At home but translocal

Nevertheless, a translocal home still has specific characteristics that make the experience of home in a translocal space different from local senses of home. Feelings of attachment and belonging are spread over more than one place, held together by manifold and heterogeneous translocal connections. These translocal connections facilitate but also demand mobility. Regular changes in location result in families, relatives, friends, or lovers being spread over different places, some always being left behind while others become (re)united. Physically distant places that are emotionally close become part of people's everyday lives as they are evoked through material objects and images. Regular visits are made to strengthen family as well as business ties, but they involve high material costs and often reach the limits of financial capacities. Moreover, having the opportunity to live and feel at home in different places also brings with it the problem of making the right decision concerning where to stay, be it for the following days, weeks, months or years. As I wanted to show in this book, the ways in which these decisions are made – when to travel, where to travel, what or whom to take on the journey – contribute to the situative, flexible, spontaneous and therefore also unpredictable nature of making home in a translocal space.

Though in academic discourses translocality is often celebrated as transgressing the constraints of the local and bringing with it new opportunities – as for example in the context of trading activities – what is often missing is a careful attention to the everyday struggles and efforts that go into sustaining translocal lives. Translocality does not always mean ease, but as the empirical examples from the Swahili context have vividly illustrated, translocality involves a number of different experiences ranging from excitement, bustle and success to frustration, disappointment, loneliness or lovesickness. Though these feelings are of course not restricted to translocal lives, they nevertheless gain a specific quality due to the translocal setting in which those living translocality are embedded but also caught.

In this book I have tried to explore the effects of mobility on conceptualisations of place and space based on close ethnographic research of Swahili trading connections. Understanding the emerging translocal space through its constitutive translocal connections allowed me to carefully examine its genuinely relational, heterogeneous and fluid character. Avoiding a celebratory overemphasis on transgression, this perspective however also opens the view for the fixed elements, stabilities and monotony that are part of the different kinds of mobilities involved in translocal connections, thus providing important counter-arguments to recent calls that argue for a radical renunciation of topography in conceptualisations of relational space. After all, living translocality is not all that different from living in more local contexts. But, providing a vivid idea of the specific character of translocality as a lived experience – with elements of ease though also with demands – has shown that it is not the same thing either. Instead, it allows for an empirically-grounded refinement of current theoretical reflections on the relationship between economy and culture as well as between mobility and space, avoiding one-sided and overly abstract conceptualisations while attending to the ways in which people actually experience and live translocality.

REFERENCES

ABOLAFIA, M. Y. (1998) Markets as cultures: An ethnographic approach, in: Callon, M. (ed.) *The law of the markets*, Blackwell: Oxford, p. 79–85.
ADEY, P. (2004) Secured and sorted mobilities: Examples from the airport, in: *Surveillance and Society* 1, p. 500–519.
—— (2006) Airports and air-mindedness: Spacing, timing and using the Liverpool Airport 1929–1939, in: *Social Cultural Geography* 7, 3, p. 343–363.
—— (2007) 'May I have your attention': airport geographies of spectatorship, position and (im)mobility. In: *Environment and Planning D: Society and Space* 25, 3, p. 515–536.
—— (2010a) *Mobility*, Routledge: London.
—— (2010b) *Aerial Life: spaces, mobilities, affects*, Wiley-Blackwell: London.
ADRIANSEN, H. & NIELSEN, T. (2005) The geography of pastoral mobility: A spatio-temporal analysis of GPS data from Sahelian Senegal, in: *GeoJournal* 64, p. 177–188.
AGIER, M. (1997) Ni trop près, ni trop loin. De l'implication ethnographique à l'engagement intellectuel, in: *Gradhiva* 21, p. 69–76.
ALPERS, E. (2009) *East Africa and the Indian Ocean*, Markus Wiener Publishers: Princeton.
AL-RASHEED, M. (2005) Transnational connections and national identity. Zanzibari Omanis in Muscat, in: Dresch, P. & Piscatori, J. P. (eds.) *Monarchies and nations: globalisation and identity in the Arab states of the Gulf*, I.B.Tauris: London, p. 96–113.
AL-RIYAMI, N. (2009) *Zanzibar: Persons and Events 1828–1972*, Dar Elhikma: London.
ALLEN, J., MASSEY, D. & PRYKE, M. (1999) *Unsettling cities: movement/settlement*, Routledge: London.
ALLEN, J., MASSEY, D. B. & COCHRANE, A. (1998) *Rethinking the region*, Routledge: London.
AMIN, A. (1998) Globalisation and regional development: A relational perspective, in: *Competition and Change* 3, p. 145–165.
—— (2002) Spatialities of globalisation, in: *Environment and Planning A* 34, p. 385–399.
—— (2004) Regions unbound: Towards a new politics of place, in: *Geografiska Annaler B* 86, p. 33–44.
AMIN, A., MASSEY, D. & THRIFT, N. (2000) *Cities for the many not the few*, Policy Press: Bristol.
AMIN, A. & THRIFT, N. (2000) What sort of economics for what sort of economic geography?, in: *Antipode* 32, p. 4–9.
—— (2002) *Cities: reimagining the urban*, Polity Press: Cambridge.
—— (2007) Cultural-economy and cities, in: *Progress in Human Geography* 31, p. 143–161.
AMSELLE, J-L. (2002) Globalization and the Future of Anthropology, in: *African Affairs* 101, p. 213–229.
ANDERSON, J. (2004) Talking whilst walking: a geographical archaeology of knowledge, in: *Area* 36, p. 254–261.
ANHEIER, H. & ISAR, Y.R. (2008) *The Cultural Economy*, Sage: London.
APPADURAI, A. (1986) Introduction: Commodities and the politics of value, in: Appadurai, A. (ed.) *The Social Life of Things*, Cambridge University Press: Cambridge, p. 3–63.
—— (1990) Disjuncture and Difference in the global cultural economy, in: *Public Culture* 2, p. 1–24.
—— (1995) The production of locality, in: Fardon, R. (ed.) *Counterworks: managing the diversity of knowledge*, p. 92–210.
—— (1996) *Modernity at large: Cultural dimensions of globalisation*, University of Minnesota Press: Minneapolis.
ARNOLD, N. (2002) Placing the shameless: Approaching poetry and the politics of Pemban-ness in Zanzibar, 1995–2001, in: *Research in African Literatures* 33, p. 140–166.
AUGÉ, M. (1995) *Non-places: Introduction to an Anthropology of Supermodernity*, Verso: London.

BÆRENHOLDT, J. et al. (2004) *Performing Tourist Places*, Ashgate: Aldershot.
BAIR, J. (2008) Analysing economic organisation: Embedded networks and global chains compared, in: *Economy and Society* 37, p. 339–364.
BAKARI, M. A. (2001) *The democratisation process in Zanzibar: A retarded transition*, GIGA-Hamburg: Hamburg.
BAKER, A. (2000) *Serious shopping: psychotherapy and consumerism*, Free Association Books: London.
BANG, A. (2008) Cosmopolitanism colonised?, in: Simpson, E. & Kresse, K. (eds.) *Struggling with history. Islam and cosmopolitanism in the Western Indian Ocean*, Columbia University Press: New York, p. 167–188.
BARNES, T. J. (2001) Retheorising economic geography: From the quantitative revolution to the 'cultural turn', in: *Annals of the Association of American Geographers* 91, p. 546–565.
—— (2005) Culture: Economy, in: Cloke, P. & Johnston, R. (eds.) *Spaces of geographical thought: Deconstructing human geography's binaries*, Sage: London, p. 61–80.
BARTELS, D. (1968) Zur wissenschaftlichen Grundlegung einer Geographie des Menschen, in: *Geographische Zeitschrift* Erdkundliches Wissen No. 19; Franz Steiner Verlag: Stuttgart.
BASCH, L., GLICK SCHILLER, N. & SZANTON BLANC, C. (1994) *Nations unbound: transnational projects, postcolonial predicaments and deterritorialised nation-states*, Routledge: New York.
BATHELT, H. & GLUECKLER, J. (2003) Toward a relational economic geography, in: *Journal of Economic Geography* 3, p. 117–144.
BATTUTA, I. (2005 [1929]) *Travels in Asia and Africa, 1325–1354*, Routledge Curzon: London.
BAUMAN, Z. (1999) *Culture as Praxis*, Sage: London.
BEAVERSTOCK, J., SMITH, R. & TAYLOR, P. (2000) World-city network: a new metageography?, in: *Annals of the Association of American Geographers* 90, p. 123–134.
BEBBINGTON, A. & KOTHARI, U. (2006) Transnational development networks, in: *Environment and Planning A* 38, p. 849–866.
BECK, R.-M. (1992) Women are Devils! A formal and stylistic analysis of *mwanameka*, in: Graebner, W. (ed.) *Sokomoko. Popular Culture in East Africa*, Rodopi: Amsterdam, p. 115–132.
BECK, U. & BECK-GERNSHEIM, E. (2009) Global Generations and the Trap of Methodological Nationalism for a Cosmopolitan Turn in the Sociology of Youth and Generation, in: *European Sociological Review* 25, p. 25–36.
BECKER, J. (1887) *La vie en Afrique*, Paris.
BELCHER, O. et al. (2008) Everywhere and nowhere: The exception and the topological challenge to geography, in: *Antipode* 40, p. 499–503.
BENDER, E. (2004) Social Lives of a Cell Phone, in: *Technology Review* [Online]. Retrieved on 26. October 2009, www.technologyreview.com/articles/print_version/wo_bender071204.asp.
BENFOUGHAL, T. (2002) Ces objets qui viennent d´ailleurs, in: Claudor-Hawad, H. (ed.) *Voyager d´un point de vue nomade*, Editions Paris-Méditerrannée: Paris, p. 133–135.
BENITEZ, J. (2006) Transnational dimensions of the digital divide among Salvadorian immigrants in the Washington DC metropolitan area, in: *Global Networks* 6, p. 181–199.
BERNARD, H. (2006) *Research methods in anthropology. Qualitative and quantitative approaches*, Sage: London.
BERNDT, C. & BOECKLER, M. (2008) Cultural geographies of economies: Performing transnational market b/orders in a global age, Working paper [Online]. Retrieved on June 2010 from http://www.geo.uni-frankfurt.de/ifh/Personen/berndt/downloads/cultural_geographies_of_economies.pdf
BHABHA, H. (1994) *The Location of Culture*, Routledge: London.
BHACKER, M. R. (1992) *Trade and empire in Muscat and Zanzibar*, Routledge: London.
BISSELL, W. (2007) Casting a long shadow: Colonial categories, cultural identities, and cosmopolitan spaces in globalizing Africa, in: *African Identities* 5, p. 181–197.

BLION, R. (2002) North of south: European immigrants' stakeholdings in southern development, in: Bryceson, D. & Vuorela, U. (eds.) *The transnational family*, Oxford: Berg, p. 231–243.
BLOMLEY, N. (1996) I'd like to dress her all over: Masculinity, power and retail space, in: Wrigley, N. & Lowe, M. (eds.) *Retailing consumption and capital*, Longman: Harlow, p. 238–256.
BLOMMAERT, J. (1999) *State Ideology and Language in Tanzania*, Rüdiger Köppe: Köln.
BLOTEVOGEL, H. (1995) Raum, in: Landesplanung, Akademie für Raumforschung und Landesplanung (ed.) *Handwörterbuch der Raumordnung*, ARL: Hannover, p. 733–740.
—— (2003) Neue Kulturgeographie – Entwicklung, Dimensionen, Potenziale und Risiken einer kulturalistischen Humangeographie, in: *Berichte zur deutschen Landeskunde* 1, p. 7–34.
BLUNT, A. et al. (2003) *Cultural geography in practice*, Oxford University Press: Oxford.
BOECKLER, M. (2004) Culture, geography and the diacritical practice of 'Oriental entrepreneurs', in: *Geographische Zeitschrift* 92, p. 39–57.
—— (2005) *Geographien kultureller Praxis. Syrische Unternehmer und die globale Moderne*, Transcript: Bielefeld.
BÖHME, H. (2005) *Topographien der Literatur. Deutsche Literatur im transnationalen Kontext*, Metzler: Stuttgart/Weimar.
BOGATU, C. (2007) *Smartcontainer als Antwort auf logistische und sicherheitsrelevante Herausforderungen in der Lieferkette*, Universitätsverlag TU Berlin: Berlin.
BOGGS, J. & RANTISI, N. (2003) The 'relational turn' in economic geography, in: *Journal of Economic Geography* 3, p. 109–116.
BOURDIEU, P. (2000) *Die zwei Gesichter der Arbeit: Interdependenzen von Zeit- und Wirtschaftsstrukturen am Beispiel einer Ethnologie der algerischen Übergangsgesellschaft*, UVK: Konstanz.
BOURDIEU, P. & WACQUANT, J. (1992) *An invitation to reflexive sociology*, University of Chicago Press: Chicago.
BOYD, M. (1989) Family and personal networks in international migration: Recent developments and new agendas, in: *International Migration Review*, p. 638–670.
BRADBURY, H. & LICHTENSTEIN, B. M. B. (2000) Relationality in organisational research: Exploring the space between, in: *Organisation Science* 11, p. 551–564.
BRAUDEL, F. (1986) *Sozialgeschichte des 15.–18.Jahrhunderts – Der Handel*, Kindler Verlag: München.
BRENNER, N. (2001) The limits of scale? Methodological reflections on scalar structuration, in: *Progress in Human Geography* 15, p. 525–548.
BRICKELL, K. & DATTA, A. (2011) Introduction: Translocal Geographies, in: Brickell, K & Datta, A: (eds.) *Translocal Geographies. Spaces, Places, Connections*, Ashgate: London, p. 3–20.
BRIDGE, G. & SMITH, A. (2003) Intimate encounters: Culture – economy – commodity, in: *Environment and Planning D: Society and Space* 21, p. 257–268.
BRODE, H. (2000) *Tippu Tip: the story of his career in Zanzibar and central Africa*, The Gallery Publications: Zanzibar.
BROMBER, K. (2006) Ustaarabu: A conceptual change in Tanganyikan newspaper discourse in the 1920s, in: Loimeier, R. & Seesemann, R. (eds.) *The Global Worlds of the Swahili*, LIT: Münster, p. 67–82.
BROWN, B. & LAURIER, E. (2005) Maps and journeys: An ethnomethodological investigation, in: *Cartographica* 4, p. 17–33.
BROWN, L. & DURRHEIM, K. (2009) Different kinds of knowing: Generating qualitative data through mobile interviewing, in: *Qualitative Inquiry* 15, p. 911–930.
BUDKE, A., KANWISCHER, D. & POTT, A. (2004) *Internetgeographien. Beobachtungen zum Verhältnis von Internet, Raum und Gesellschaft*, Franz Steiner Verlag: Stuttgart.
BUNNELL, T. (2007) Post-maritime transnationalisation: Malay seafarers in Liverpool, in: *Global Networks* 7, p. 412–429.
BURAWOY, M. (2000) *Global ethnography: Forces, connections, and imaginations in a postmodern world*, University of California Press: Berkeley.

CALLON, M. (1986) Some elements of a sociology of translation: Domestication of the scallops and the fishermen of Saint Brieuc Bay, in: Law, J. (ed.) *Power, action and belief: A new sociology of knowledge?*, Routledge: London, p. 196–233.
—— (1998) *The laws of the markets*, Blackwell: Oxford.
—— (1999) Actor-Network-Theory: The market test, in: Law, J. & Hassard, J. (eds.) *Actor Network Theory and after*, Blackwell: Oxford, p. 181–195.
CAMERON, G. (2004) Political violence, ethnicity and the agrarian question in Zanzibar, in: Caplan, P. & Topan, F. (eds.) *Swahili Modernities: Culture, Politics and Identity on the East Coast of Africa*, Africa World Press: Trenton, NJ, p. 103–119.
CANZLER, W., KAUFMANN, V. & KESSELRING, S. (2008) *Tracing mobilities: Towards a cosmopolitan perspective*, Ashgate: London.
CAPLAN, P. (2004) Introduction, in: Caplan, P. & Topan, F. (eds.) *Swahili modernities: Culture, politics and identity on the East Coast of Africa*, Africa World Press: Trenton, NJ, p. 1–18.
CAPLAN, P. & TOPAN, F. (2004) *Swahili modernities: Culture, Politics and Identity on the East Coast* Africa World Press: Trenton, NJ.
CARPIANO, R. (2009) Come take a walk with me: The "go-along" interview as a novel method for studying the implications of place for health and well-being, in: *Health and Place* 15, p. 263–272.
CARTER, T. & DUNSTON, L. (2007) *Dubai*, Mairdumont: Ostfildern.
CASEY, E. (1993) *Getting back into place*, Indiana University Press:
CASSIRER, E. (1953) *Substance and function*, Dover: New York.
CASSON, L. (1989) *The Periplus Maris Erythraeai*, Princeton University Press: Princeton.
CASTELLS, M. (1996) *The rise of the network society*, Blackwell: Oxford.
—— (1997) *The power of identity*, Blackwell: Oxford.
—— (1998) *End of millenium*, Blackwell: Oxford.
CASTLES, S. (2008) The factors that make and unmake migration policies, in: Portes, A. & DeWind, J. (eds.) *Rethinking migration: new theoretical and empirical perspectives*, Berghahn Books: New York, p. 29–61.
—— (2009) Development and migration – Migration and development: What comes first? Global perspective and African experiences, in: *Theoria* 56, p. 1–31.
CASTREE, N. (2003) Environmental issues: Relational ontologies and hybrid politics, in: *Progress in Human Geography* 27, p. 203–211.
—— (2004) Economy and culture are dead! Long live economy and culture!, in: *Progress in Human Geography* 28, p. 204–226.
CHAMI, F. A. & MSEMWA, P. J. (1997) A New Look at Culture and Trade on the Azanian Coast, in: *Current Anthropology* 38, p. 673–677.
CHAUDHURI, K. N. (1985) *Trade and civilisation in the Indian Ocean*, Cambridge University Press: Cambridge.
CHELPI-DEN HAMER, M. & MAZZUCATO, V. (2010) The role of support networks in the initial stages of integration: The case of West African newcomers in the Netherlands, in: *International Migration* 48, p. 31–57.
CHERNILO, D. (2007) *A social theory of the nation-state: The political forms of modernity beyond methodological nationalism*, Routledge: London.
—— (2010) Methodological nationalism and the domestic analogy: classical resources for their critique, in: *Cambridge Review of International Affairs* 23, p. 87–106.
CHRISTALLER, W. (1933) *Die Zentralen Orte in Süddeutschland. Eine ökonomisch-geographische Untersuchung über die Gesetzmäßigkeiten der Verbreitung und Entwicklung der Siedlungen mit städtischen Funktionen*, Fischer: Jena.
CLAYTON, A. (1981) *The Zanzibar revolution and its aftermath*, C. Hurst & Co: London.
CLIFFORD, J. (1983) On Ethnographic Authority, in: *Representations* 1, p. 118–146.
—— (1986) Introduction: Partial truths, in: Clifford, J. & Marcus, G. (eds.) *Writing culture: the poetics and politics of ethnography*, University of California Press: Berkeley, p. 1–26.

—— (1997) *Routes: Travel and translation in the late twentieth century*, Harvard University Press: Cambridge/Mass.
CLIFFORD, N., FRENCH, S. and VALENTINE, G. (2010) *Key Methods in Geography*, Sage: London.
CLOKE, P. et al. (2004) *Practicing Human Geography*, Sage: London.
COLLINS, F. L. (2010) Negotiating un/familiar embodiments: Investigating the corporeal dimensions of South Korean international student mobilities in Auckland, New Zealand, in: *Population, Space and Place* 16, p. 51–62.
COLMAN, F. (2005) Rhizome, in: Parr, A. (ed.) *The Deleuze Dictionary*, Edinburgh University Press: Edinburgh, p. 231–235.
COMAROFF, J. & COMAROFF, J. (2003) Ethnography on an awkward scale: Postcolonial anthropology and the violence of abstraction, in: *Ethnography* 4, p. 147–179.
CONRADSON, D. & LATHAM, A. (2005) Transnational urbanism: Attending to everyday practices and mobilities, in: *Journal of Ethnic and Migration Studies* 31, p. 227–233.
CONSTANTIN, F. & LE GUENNEC-COPPENS, F. (1988) Dubaï Street, Zanzibar... in: *Politique africaine* 30, p. 7–21.
COOK, I. et al (2004) Follow the Thing: Papaya, in: *Antipode* 36, p. 642–664.
—— (2005) Positionality / Situated knowledge, in: Atkinson, D. et al. (eds.) *Critical concepts in cultural geography*, IB Tauris: London, p. 16–26.
—— (2006) Geographies of food: Following, in: *Progress in Human Geography* 30 (5), p. 655–666.
COOK, I. & CRANG, P. (1996) The world on a plate: culinary culture, displacement and geographical knowledges, in: *Journal of Material Culture* 1, p. 131–153.
COOK, I. & HARRISON, M. (2007) Follow the Thing: West Indian Hot Pepper Sauce, in: *Space and Culture* 10 (1), p. 40–63.
COOPER, F. (1977) *Plantation slavery on the east coast of Africa*, Yale University Press: New Halen.
COPE, M. (2010) A history of qualitative research in geography, in: DeLyser, D., Herbert, S., Aitken, S., Crang, M. & McDowell, L. (eds.) *The SAGE Handbook of Qualitative Geography*, Sage: London, p. 25–45.
COSGROVE, D. (1983) Towards a radical cultural geography, in: *Antipode* 15, p. 1–11.
COUCLELIS, H. (2009) Rethinking time geography in the information age, in: *Environment and Planning A* 41, p. 1556–1575.
COX, K. (2005) Global: Local, in: Cloke, P. & Johnston, R. (eds.) *Spaces of geographical thought: deconstructing human geography's binaries*, Sage: London, p. 175–198.
CRANG, M. (2005) Qualitative methods: there is nothing outside the text?, in: *Progress in Human Geography* 29, p. 225–233.
CRANG, M. & COOK, I. (2007) *Doing ethnographies*, Sage: London.
CRANG, M., CRANG, P. & MAY, J. (1999) *Virtual Geographies*, Routledge: London.
CRANG, P. (1994) It's showtime: on the workplace geographies of display in a restaurant in southeast England, in: *Environment and Planning D: Society and Space* 12, p. 675–704.
—— (1997) Cultural turns and the (re)constitution of economic geography, in: Lee, R. & Willis, J. (eds.) *Geographies of Economies*, Arnold: London, p. 3–15.
—— (2010) Cultural geography: after a fashion, in: *Cultural Geographies* 17, p. 191–201.
CRANG, P. & ASHMORE, S. (2009) The transnational spaces of things: Asian textiles in Britain and the grammar of ornament, in: *European Review of History* 16, p. 655–678.
CRANG, P. & JACKSON, P. (2001) Geographies of consumption, in: Morley, D. & Robins, K. (eds.) *British cultural studies: Geography, nationality, and identity*, Oxford University Press: Oxford, p. 327–342.
CRESSWELL, T. (2004) *Place. A short introduction*, Blackwell: Oxford.
—— (2006) *On the move*, Routledge: London.

—— (2008) Understanding mobility holistically: The case of hurricane Katrina, in: Bergmann, S. & Sager, T. (eds.) *The ethics of mobilities: Rethinking place, exclusion, freedom and environment*, Ashgate: London, p. 129–142.
—— (2010) New cultural geography – an unfinished project?, in: *cultural geographies* 17, p. 169–174.
CRESSWELL, T. & MERRIMAN, P. (2010) *Geographies of mobilities: Practices, spaces, subjects*, Ashgate: London.
CREWE, L. (2000) Geographies of retailing and consumption, in: *Progress in Human Geography* 24, p. 275–290.
CREWE, L. & GREGSON, N. (1998) Tales of the unexpected: exploring car boot sales as marginal spaces of consumption, in: *Transactions of the Institute of British Geographers* 23, p. 39–53.
CROUCH, D. (2010) Flirting with space: thinking landscape relationally, in: *Cultural Geographies* 17, p. 5–18.
CUNNINGHAM, S., BANKS, J. & POTTS, J. (2008) The cultural economy: the shape of the field, in: Anheier, H. & Isar, Y.R. (2008) *The Cultural Economy*, Sage: London, p. 15–27.
CURTIN, P. D. (1984) *Cross-cultural Trade in World History*, Cambridge University Press: Cambridge.
DAVIES, G. & DWYER, C. (2007) Qualitative methods: Are you enchanted or are you alienated?, in: *Progress in Human Geography* 31, p. 257–266.
DE BRUIJN, M., NYAMNJOH, F. & BRINKMAN, I. (2009) *Mobile phones: The new talking drums in everyday Africa*, Langaa and African Studies Centre: Cameroon and Leiden.
DE VERE ALLEN, J. (1993) *Swahili origins: Swahili culture and the Shungwaya phenomenon*, James Currey: Oxford.
DELEUZE, G. (1988) *Le pli: Leibniz et le baroque*, Minuit: Paris.
—— (1995) *Negotiations*, Colombia University Press: New York.
DELEUZE, G. & GUATTARI, F. (1976) *Rhizome*, Minuit: Paris.
—— (1977) *Rhizom*, Merve: Berlin.
—— (1980) *Mille plateaux*, Minuit: Paris.
DELYSER, D., HERBERT, S., AITKEN, S., CRANG, M. and MCDOWELL, L. (2010) *The SAGE Handbook of Qualitative Geography*, Sage: London.
DESCARTES, R. (1986 [1641]) *Meditationes de prima philosophia*, Reclam: Stuttgart.
DICKEN, P. et al. (2001) Chains and networks, territories and scales: towards a relational framework for analysing the global economy, in: *Global Networks* 1, p. 89–112.
DICKEN, P. & MALMBERG, A. (2001) Firms in territories: a relational perspective, in: *Economic Geography* 77, p. 345–363.
DOEL, M. (1999) *Poststructuralist geographies. The diabolical art of spatial science*, Edinburgh University Press: Edinburgh.
DÖRING, J. & THIELMANN, T. (2008) *Spatial turn. Das Raumparadigma in den Kultur- und Sozialwissenschaften*, Transcript: Bielefeld.
DUGAY, P. & PRYKE, M. (2002) *Cultural economy – cultural analysis and commercial life*, SAGE: London.
DUNCAN, J. (1980) The superorganic in American cultural geography, in: *Annals of the Association of American Geographers* 70, p. 181–198.
DUNCAN, J. & DUNCAN, N. (1996) Reconceptualising the idea of culture in geography: a reply to Don Mitchell, in: *Transactions of the Institute of British Geographers* 21, p. 576–579.
DWYER, C. & JACKSON, P. (2003) Commodifying difference: Selling EASTern Fashion, in: *Environment and Planning D: Society and Space* 21, p. 269–291.
DYCK, I. (2005) Feminist geography, the 'everyday' and local-global relations, in: *Canadian Geographer/Le Geographe canadien* 49, p. 233–243.
EARMAN, J. (1989) *World enough and space-time: absolute versus relational theories of space and time*, MIT Press: Cambridge.
EASTMAN, C. (1971) Who are the Waswahili?, in: *Africa* 41, p. 228–236.

EDENSOR, T. (2002) *National Identity, popular culture and everyday life*, Berg: Oxford.
EINSTEIN, A. (1960) Vorwort, in: Jammer, M. (ed.) *Das Problem des Raumes. Die Entwicklung der Raumtheorien*, Wissenschaftliche Buchgesellschaft: Darmstadt, p. XII–XVII.
ELIAS, N. (1976) *Über den Prozess der Zivilisation*, Suhrkamp: Frankfurt am Main.
—— (1978) *Was ist Soziologie?*, Juventa: München.
ELSHESHTAWY, Y. (2008) Transitory sites: Mapping Dubai's 'forgotten' urban spaces, in: *International Journal of Urban and Regional Research* 32, p. 968–988.
EMIRBAYER, M. (1997) Manifesto for a relational sociology, in: *The American Journal of Sociology* 103, p. 281–317.
EMIRBAYER, M. & GOODWIN, J. (1994) Network analysis, culture, and the problem of agency, in: *The American Journal of Sociology* 99, p. 1411–1454.
ENGLAND, K. (1994) Getting personal: reflexivity, positionality and feminist research, in: *The Professional Geographer* 46, p. 80–89.
ENGSTROEM, Y. & BLACKLER, F. (2005) On the Life of the Object, in: *Organization* 12, p. 307–330.
ESCOBAR, A. (1994) Welcome to cyberia: notes on the anthropology of cyberculture, in: *Current Anthropology* 35, p. 56–76.
FEATHERSTONE, D., PHILLIPS, R. & WATERS, J. (2007) Introduction: spatialities of transnational networks, in: *Global Networks* 7, p. 383–391.
FELTER, H.W. & LLOYD, J.U. (1898) *King's American Dispensatory*, Ohio Valley Co.: Cincinatti.
FIELDING, T. (1992) Migration and culture, in: Champion, T. & Fielding, T. (eds.) *Migration processes and patterns. Research processes and prospects*, Belhaven Press: London, p. 201–212.
FORTIER, A.-M. (1999) Re-membering places and the performance of belonging(s), in: *Theory Culture and Society* 16, p. 41–64.
FOWLER, C. (2005) Reexploring transport geography and networks: A case study of container shipments to the West Coast for the United States, in: *Environment and Planning A* 38, p. 1429–1448.
FREEMAN-GRENVILLE, G. (1960) East African coin finds and their historical significance, in: *Journal of African History* 1, p. 31–43.
FREIDBERG, S. (2001) On the trail of the global green bean: Methodological considerations in multi-site ethnography, in: *Global Networks* 1, p. 353–368.
FREITAG, U. & VON OPPEN, A. (2010a) *Translocality. The study of globalising processes from a southern perspective*, Brill: Leiden.
—— (2010b) 'Translocality': An approach to connection and transfer in regional studies, in: Freitag, U. & von Oppen, A. (eds.) *Translocality. The study of globalising processes from a southern perspective*, Brill: Leiden, p. 1–24.
FRESNOZA-FLOT, A. (2009) Migration status and transnational mothering: the case of Filipino migrants in France, in: *Global Networks* 9, p. 252–270.
GADAMER, H.-G. (1960) *Wahrheit und Methode. Grundzüge einer philosophischen Hermeneutik*. J.V.B. Mohr (Paul Siebeck): Tübingen.
—— (1975) *Truth and Method*, Scheed and Ward: London.
GEBHARDT, H., REUBER, P. & WOLKERSDORFER, G. (2003) *Kulturgeographie. Aktuelle Ansätze und Entwicklungen*, Spektrum: Heidelberg.
GEERTZ, C. (1963) *Peddlers and princes. Social change and economic modernisation in two Indonesian towns*, University of Chicago Press: Chicago.
—— (1973) Thick description: Toward an interpretive theory of culture, in: Geertz, C. (ed.) *The interpretation of cultures. Selected Essays*, Basic Books: New York, p. 3–30.
GERMANN MOLZ, J. (2006) Watch us wander: mobile surveillance and the surveillance of mobility, in: *Environment and Planning A* 38, p. 377–393.
—— (2007) Cosmopolitans on the couch: Mobile hospitality and the internet, in: Molz, J. & Gibson, S. (eds.) *Mobilising hospitality: the ethics of social relations in a mobile world*, Ashgate: Aldershot, p. 65–102.

—— (2008) Global abode: Home and mobility in narratives of round-the-world travel, in: *Space and Culture* 11, p. 325–342.
GIBSON, C. & KONG, L. (2005) Cultural Economy: A critical review, in: *Progress in Human Geography* 29, p. 541–561.
GILBERT, E. (2004) *Dhows and the colonial economy of Zanzibar, 1860–1970*, James Curry: London.
—— (2007) Oman and Zanzibar: The historical roots of a global community, in: Prabha Ray, H. & Alpers, E. A. (eds.) *Cross currents and community networks. The history of the Indian Ocean world*, Oxford University Press: Oxford, p. 163–178.
GILMORE, Z. (2007) *Footprint Dubai*, Footprint Travel Guides: Bath.
GLASSMAN, J. (2004) Slower than a massacre: the multiple sources of racial thought in colonial Africa, in: *American Historical Review* 109, p. 720–754.
—— (2011). *War of words, war of stones*. Indiana University Press: Bloomington.
GLICK SCHILLER, N., BASCH, L. & BLANC-SZANTON, C. (1992) *Towards a transnational perspective on migration: Race, class, ethnicity and nationalism reconsidered*, New York Academy of Science: New York.
GOLDRING, L. (1999) Power and status in transnational social spaces, in: Pries, L. (ed.) *Migration and transnational social spaces*, Ashgate: Aldershot, p. 162–186.
GOSS, J. (2004) Geographies of Consumption I, in: *Progress in Human Geography* 28, p. 369–380.
GRABHER, G. (2006) Trading routes, bypasses, and risky intersections: mapping the travels of networks between economic sociology and economic geography, in: *Progress in Human Geography* 30, p. 163–189.
GRAHAM, S. (1998) The end of geography or the explosion of place? Conceptualizing space, place and information technology, in: *Progress in Human Geography* 22, p. 165–185.
GRAHAM, S. & HEALEY, P. (1999) Relational concepts of space and place: Issues for planning theory and practice, in: *European Planning Studies* 7, p. 623–646.
GRANOVETTER, M. (1973) The strength of weak ties, in: *American Journal of Sociology* 78, p. 1360–1380.
—— (1983) The strength of weak ties: a network theory revisited, in: *Sociological Theory* 1, p. 201–233.
—— (1985) Economic action and social structure: The problem of embeddedness, in: *American Journal of Sociology* 91, p. 481–510.
GREGORY, D. (1995) Imaginative Geographies, in: *Progress in Human Geography* 19, p. 447–485.
GREGSON, N. & CREWE, L. (1994) Beyond the high street and the mall: car boot fairs and the new geographies of consumption in the 1990s, in Area 226, p. 261–267.
GREGSON, N., SIMONSEN, K. and VAIOU, D. (2001) Whose economy for whose culture?, in: *Antipode* 33, p. 616–646.
GRILLO, R. & RICCIO, B. (2004) Translocal Development: Italy-Senegal, in: *Population, Space and Place* 10, p. 99–111.
GUARNIZO, L. & SMITH, M. (1998) The locations of transnationalism, in: Smith, M. & Guarnizo, L. (eds.) *Transnationalism from below*, Transnation Publisher: New Brunswick, p. 3–31.
GUPTA, A. & FERGUSON, J. (1997) Discipline and practice: 'The field" as site, method, and location in anthropology, in: Gupta, A. & Ferguson, J. (eds.) *Anthropological Locations – boundaries and grounds of a field science*, University of California Press: Berkeley, California, p. 1–46.
GURNAH, A. (2002) *By the Sea*, Bloomsbury: London.
—— (2005) *Desertion*, Bloomsbury: London.
HÄGERSTRAND, T. (1967) *Innovation diffusion as a spatial process*, University of Chicago Press: Chicago.
—— (1970) What about people in regional science, in: *Regional Science Association* 24, p. 7–21.

—— (1985) Time-geography: focus on the corporeality of man, society and environment, in: Aida, S. (ed.) *The Science and Praxis of Complexity*, United Nations University: Tokyo, p. 193–216.
HAGGETT, P. (1965) *Locational analysis in human geography*, Edward Arnold: London.
HAGGETT, P. & Chorley, R. (1969) *Network analysis in geography*, Edward Arnold: London.
HAHNER-HERZOG, I. (1990) *Tippu Tip und der Elfenbeinhandel in Ost- und Zentralafrika im 19. Jahrhundert*, tuduv: München.
HALL, S. (1996) When was the "post-colonial"? Thinking at the limit, in: Chambers, L. & Curti, L. (eds.) *The postcolonial question. Common skies, divided horizons*, Routledge: London, p. 242–260.
HALL, T., LASHUA, B. and COFFEY, A. (2008) Sound and the everyday in qualitative research, in: *Qualitative Inquiry* 14, p. 1019–1040.
HANNERZ, U. (1992) *Cultural complexity. Studies in the social organisation of meaning*, Columbia University Press: New York.
—— (1996) *Transnational connections*, Routledge: London.
—— (2003) Being there . . . and there . . . and there! Reflections on multi-site ethnography, in: *Ethnography* 4 p. 201–216.
HARD, G. (1993) Über Räume reden. Zum Gebrauch des Wortes 'Raum' in sozialwissenschaftlichem Zusammenhang, in: Maier, J. (ed.) *Die aufgeräumte Welt – Raumbilder und Raumkonzepte in Zeitalter globaler Marktwirtschaft*, Evangelische Akademie Loccum: Rehburg-Loccum, p. 53–77.
—— (2008) Der spatial turn, von der Geographie her beobachtet, in: Döring, J. &. Thielman, T. (eds.) *Spatial turn. Das Raumparadigma in den Kultur- und Sozialwissenschaften*, Transcript: Bielefeld, p. 263–316.
HARRIS, P. (2005) The smooth operator: Serres prolongs Poe, in: Abbas, N. (ed.) *Mapping Michel Serres*, University of Michigan Press: Ann Arbor, p. 113–135.
HART, G. (2004) Geography and development: critical ethnographies, in: *Progress in Human Geography* 28, p. 91–100.
HARVEY, D. (1969) *Explanation in geography*, Edward Arnold: London.
—— (1996) *Justice, nature and the geography of difference*, Blackwell: Oxford.
HÄUSSLING, R. (2009) *Grenzen von Netzwerken*, VS Verlag: Wiesbaden.
HAY, A. (2000) Transport geography, in: Johnston, R., Gregory, D. & Smith, D. (eds.) *The Dictionary of Human Geography*, Blackwell: Oxford, p. 855–856.
HEIDEGGER, M. (1927) *Sein und Zeit*, Max Niemeyer: Tübingen.
HENAFF, M. (2005) Of stones, angels, and humans – Michel Serres and the global city, in: Abbas, N. (ed.) *Mapping Michel Serres*, University of Michigan Press: Michigan, p. 170–189.
HERBERT, S. (2000) For ethnography, in: *Progress in Human Geography* 24, p. 550–568.
HESS, M. (2004) 'Spatial' relationships? Towards a reconceptualisation of embeddedness, in: *Progress in Human Geography* 28, p. 165–186.
HETHERINGTON, K. (1997) *The badlands of modernity: heterotopia and social ordering*, Hetherington: London.
HEUMAN, J. (2003) Beyond political economy versus cultural studies? The new 'cultural economy', in: *Journal of Communication Inquiry* 27, p. 104–109.
HEYDENREICH, S. (2000) *Aktionsräume in dispersen Stadtregionen*, L.I.S. Verlag: Passau.
HILL, J. (2007) The story of the amulet, in: *Journal of Material Culture* 12 (1), p. 65–87.
HINCHLIFFE, S. et al. (2005) Urban wild things: a cosmopolitical experiment, in: *Environment and Planning D* 23, p. 643–658.
HINE, C. (2005) *Virtual Methods: Issues in Social Research on the Internet*, Berg: Oxford.
HITCHINGS, R. & JONES, V. (2004) Living with plants and the exploration of botanical encounter within human geographic research practice, in: *Ethics, Place and Environment* 7, p. 3–18.
HO, E. L.-E. (2011) Migration trajectories of 'highly skilled' middling transnationals: Singaporean transmigrants in London, in: *Population, Space and Place* 17, p. 116–129.

HO, E. (2006) The Graves of Tarim. Genealogy and Mobility across the Indian Ocean, University of California Press: Los Angeles.
HOLLIFIELD, J. (2008) The emerging migration state, in: Portes, A. & DeWind, J. (eds.) *Rethinking migration: new theoretical and empirical perspective*, Berghahn Books: New York, p. 62–89.
HOOKS, b. (1992) *Black looks: Race and representation*, South End Press: Boston.
HORST, H. A. (2006) The blessings and burdens of communication: cell phones in Jamaican transnational social fields, in: *Global Networks* 6, p. 143–159.
HORTON, F. & REYNOLDS, D. (1969) An investigation of individual action spaces: A progress report, in: *Annals of the Association of American Geographers* 1, p. 70–75.
HORTON, M. (1987) The Swahili corridor, in: *Scientific American* 257, p. 86–93.
HORTON, M. & MIDDLETON, J. (2000) *The Swahili: the social landscape of a mercantile society*, Blackwell: Oxford.
HUDSON, R. (2004) Conceptualizing economies and their geographies: spaces, flows and circuits, in: *Progress in Human Geography* 28, p. 447–471.
HUGHES, A. (2007) Geographies of exchange and circulation: flows and networks of knowledgeable capitalism, in: *Progress in Human Geography* 31, p. 527–535.
HULME, M. & TRUCH, A. (2006) The role of interspace in sustaining identity, in: *Knowledge, Technology & Policy* 19, p. 45–53.
HURST, E. (1974) *Transportation geography: comments and readings*, McGraw-Hill: New York.
ILIFFE, J. (1979) *A modern history of Tanganyika*, Cambridge University Press: Cambridge.
INGOLD, T. (2006) Rethinking the animate, re-animating thought, in: *Ethnos* 71, p. 9–20.
— (2007) *Lines: a brief history*, Routledge: London.
INKPEN, R., COLLIER, P. & RILEY, M. (2007) Topographic relations: developing a heuristic device for conceptualising networked relations, in: *Area* 39, p. 536–543.
IVANOV, P. (2010) Verschleierung als Praxis. Gedanken zur Beziehung zwischen Person, Gesellschaft und materieller Welt in Sansibar, in: Tietmeyer, E. et al. (eds.) Die Sprache der Dinge – Kulturwissenschaftliche Perspektiven auf die materielle Kultur, Waxmann: Münster p. 135–148.
JACKSON, P. (1980) A plea for cultural geography, in: *Area* 12, p. 110–113.
— (1996) The idea of culture: a response to Don Mitchell, in: *Transactions of the Institute of British Geographers* 21, p. 572–585.
— (2002) Commercial cultures: transcending the cultural and the economic, in: *Progress in Human Geography* 26, p. 3–18.
JACKSON, P., CRANG, P. & DWYER, C. (2004) Introduction: The spaces of transnationality, in: Jackson, P., Crang, P. & Dwyer, C. (eds.) *Transnational spaces*, Routledge: London, p. 1–23.
JACKSON, S. & MOORES, S. (1995) *The politics of domestic consumption: critical readings*, Harvester Wheatsheaf: London.
JAZEEL, T. (2003) Unpicking Sri Lankan 'island-ness' in Romesh Gunesekera's 'Reef', in: *Journal of Historical Geography* 29, p. 582–598.
JESSOP, B., BRENNER, N. & JONES, M. (2008) Theorising sociospatial relations, in: *Environment and Planning D: Society and Space* 26, p. 389–401.
JESSOP, B. & OOSTERLYNCK, S. (2008) Cultural political economy: On making the cultural turn without falling into soft economic sociology, in: *Geoforum* 39, p. 1155–1169.
JONES, A. (2008) Beyond embeddedness: economic practices and the invisible dimensions of transnational business activity, in: *Progress in Human Geography* 32, p. 71–88.
JONES, M. (2009) Phase space: geography, relational thinking, and beyond, in: *Progress in Human Geography* 33, p. 487–506.
KAPLAN, C. (2006) Mobility at war: The cosmic view of US 'Air Power', in: *Environment and Planning A* 38, p. 395–407.
KASESNIEMI, E.-L. & RAUTIAINEN, P. (2002) Mobile culture of children and teenagers in Finland, in: Katz, J. & Aakhus, M. (eds.) *Perpetual contact. Mobile communication, private talk, public performance*, Cambridge University Press: Cambridge, p. 170–192.

KATZ, J. & SUGIYAMA, S. (2006) Mobile phones as fashion statements: Evidence from student surveys in the US and Japan, in: *New Media Society* 8 (2), p. 321–337.
KESSELRING, S. (2009) Global transfer points: the making of airports in the mobile risk society, in: Cwerner, S., Kesselring, S. & Urry, J. (eds.) *Aeromobilities*, Routledge: London, p. 39–60.
KHAMIS, S. A. M. (2000) The Heterogeneity of Swahili Literature, in: *Nordic Journal of African Studies* 9, p. 11–21.
–––– (2004) Versatility of the Taarab lyric: Local aspects and global influences, in: *Swahili Forum* 11, p. 3–37.
KILDUFF, M. & TSAI, W. (2003) *Social networks and organizations*, Sage: London.
KINDON, S. (2003) Participatory video in geographic research: a feminist practice of looking?, in: *Area* 35, p. 142–153.
KINDY, H. (1972) *Life and politics in Mombasa*, East African Publishing House: Nairobi.
KOPYTOFF, I. (1986) The cultural biography of things: Commoditization as process, in: Appadurai, A. (ed.) *The Social Life of Things*, Cambridge University Press: Cambridge, p. 64–91.
KRACAUER, S. (1995) Georg Simmel, in: Kracauer, S. & Levin, T. (eds.) *The Mass Ornament. Weimar Essays*, Harvard University Press: Cambridge, p. 225–258.
KRAIDY, M. (2005) *Hybridity. Or the Cultural Logic of Globalization*. Temple University Press: Philadelphia.
KRAMER, C. (2005) *Zeit für Mobilität*, Franz Steiner Verlag: Stuttgart.
KRESSE, K. (2007) *Philosophising in Mombasa: knowledge, Islam and intellectual practice on the Swahili coast*, Edinburgh University Press for the International African Institute: Edinburgh.
KUSENBACH, M. (2003) Street phenomenology: The go-along as ethnographic research tool, in: *Ethnography* 4, p. 455–485.
KWAN, M.-P. (2004) GIS methods in time-geographic research: Geocomputation and geovisualisation of human activity patterns, in: *Geografiska Annaler* 86, p. 267–280.
LACHENMANN, G. (2008) Transnationalisation, translocal spaces, gender and development – methodological challenges, in: Remus, G. A., Gerharz, E. & Rescher, G. (eds.) *The making of world society: perspectives from transnational research*, Transcript: Bielefeld, p. 51–74.
LARSEN, J., AXHAUSEN, K. W. & URRY, J. (2006) Geographies of social networks: Meetings, travel and communications, in: *Mobilities* 1, p. 261–283.
LARSEN, K. (2004) Change, continuity and contestation: The politics of modern identities in Zanzibar, in: Caplan, P. & Topan, F. (eds.) *Swahili modernities*, Africa World Press: Trenton NJ, p. 121–144.
–––– (2008) *Where humans and spirits meet: the politics of rituals and identified spirits in Zanzibar*, Berghahn Books: New York.
LASH, S. & URRY, J. (1994) *Economies of Signs & Spaces*, Sage: London.
LATHAM, A. (2002) Re-theorising the scale of globalisation: Topologies, actor-networks, and cosmopolites, in: Herod, A. & Wright, M. (eds.) *The geography of power: Making scale*, Blackwell: Oxford, p. 115–144.
–––– (2003) Research, performance, and doing human geography: some reflections on the diary-photograph, diary-interview method, in: *Environment and Planning A* 35, p. 1993–2017.
LATOUR, B. (1986) The powers of association, in: Law, J. (ed.) *Power, action, belief*, Routledge: London, p. 264–280.
–––– (1987) *Science in action*, Open University Press: Milton Keynes.
–––– (1988) A relativistic account of Einstein's relativity, in: *Social Studies of Science* 18, p. 3–44.
–––– (1993) *We have never been modern*, Harvard University Press: Cambridge.
–––– (1996) *Aramis, or, the love of technology*, Harvard University Press: Cambridge, Massachusetts.
–––– (1999) On recalling ANT, in: Law, J. & Hassard, J. (eds.) *Actor-Network-Theory and after*, Blackwell: Oxford, p. 15–25.
–––– (2005) *Reassembling the social. An introduction to Actor-Network-Theory*, Oxford University Press: New York.

LAURIER, E. (2004) Doing office work on the motorway, in: *Theory Culture and Society* 21, p. 261–278.
LAURIER, E. & PHILO, C. (1999) X-morphising: review essay of Bruno Latour's Aramis, or the love of technology, in: *Environment and Planning A* 31, p. 1047–1071.
—— (2006) Possible geographies: a passing encounter in a cafe, in: *Area* 38, p. 353–363.
—— (2007) A parcel of muddling muckworms: revisiting Habermas and the English Coffee House, in: *Social and Cultural Geography* 8, p. 259–281.
LAW, J. (1992) Notes on the theory of the actor network – Ordering, strategy and heterogeneity, in: *Systems Practice* 5, p. 379–393.
—— (1994) *Organising modernity*, Blackwell: Oxford.
—— (1999) After ANT: complexity, naming and topology, in: Law, J. & Hassard, J. (eds.) *Actor network theory and after*, Blackwell: Oxford, p. 1–14.
—— (2002) Economics as interference, in: DuGay, P. & Pryke, M. (eds.) *Cultural economy – cultural analysis and commercial life*, SAGE: London, p. 21–38.
—— (2004) *After method. Mess in social science research*, Routledge: London.
—— (2006 [1997]) Traduction/trahison: Notes on ANT, in: *Convergencia* 13, p. 47–72.
LAW, J. & HASSARD, J. (1999) *Actor-network theory and after*, Blackwell: Oxford.
LAW, L. (2001) Home cooking: Filipino women and geographies of the senses in Hong Kong, in: *Cultural Geographies* 8, p. 264–283.
LE COUR GRANDMAISON, C. (1989) Rich cousins, poor cousins: Hidden stratification among the Omani Arabs in Eastern Africa, in: *Africa* 59 (2), p. 176–184.
LE GUENNEC-COPPENS, F. (2002) Les swahili: Une singularité anthropologique en afrique de l'est, in: *Journal des africanistes* 72, p. 5–70.
LE GUENNEC-COPPENS, F. & CAPLAN, P. (1991) *Les swahili entre afrique et arabie*, Karthala: Paris.
LEADER-WILLIAMS, N., KAYERA, J. & OVERTON, G. (1996) *Mining in protected areas in Tanzania*, IIED: London.
LEE, J. & INGOLD, T. (2006) Fieldwork on foot: perceiving, routing and socialising, in: Coleman, S. & Collins, P. (eds.) *Locating the field: space, place and context in anthropology*, Berg: Oxford, p. 67–86.
LEES, L. (2003) Urban geography: 'New' urban geography and the ethnographic void, in: *Progress in Human Geography* 27, p. 107–113.
LEFEBVRE, H. (1968) *Le droit a la ville*, Anthropos: Paris.
LEIBNIZ, G. (1966 [1904]) *Hauptschriften zur Grundlegung der Philosophie*, Felix Meiner Verlag: Hamburg.
LEISS, W., KLINE, S. & JHALLY, S. (1990) *Social communication in advertising*, Routledge: London.
LEITNER, H. et al. (2002a) *'The city is dead, long live the net': Harnessing European interurban networks for a neoliberal agenda*, Blackwell Publishers: Oxford.
LEITNER, H., PAVLIK, C. & SHEPPARD, E. (2002b) Networks, governance, and the politics of scale: inter-urban networks and the European Union, in: Herod, A. & Wright, M. W. (eds.) *Geographies of power: placing scale*, Blackwell: Malden, p. 274–303.
LESLIE, D. & REIMER, S. (1999) Spatializing commodity chains, in: *Progress in Human Geography* 23, p. 401–420.
LEY, D. (2004) Transnational spaces and everyday lives, in: *Transactions of the Institute of British Geographers* 29, p. 151–164.
LEY, D. & OLDS, K. (1988) Landscape as spectacle: world's fairs and the culture of heroic consumption, in: *Environment and Planning D: Society and Space* 6, p. 191–212.
LI, W. (1998) Anatomy of a new ethnic settlement: the Chinese ethnoburb in Los Angeles, in: *Urban Studies* 35, p. 479–501.

LICOPPE, C. (2004) 'Connected' presence: the emergence of a new repertoire for managing social relationships in a changing communication technoscape, in: *Environment and Planning D: Society and Space* 22, p. 135–156.
LIMB, M. & DWYER, C. (2001) *Qualitative Methodologies for Geographers*, Arnold: London.
LOFCHIE, M. (1965) *Zanzibar: Background to the revolution*, Princeton University Press: Princeton.
LOFLAND, J. (1995) Analytic ethnography: features, failings and futures, in: *Journal of Contemporary Ethnography* 24, p. 30–67.
LOSSAU, J. (2008) Kulturgeographie als Perspektive. Zur Debatte um den cultural turn in der Humangeographie – eine Zwischenbilanz. In: *Berichte zur deutschen Landeskunde* 82, p. 317–343.
LOSSAU, J. & LIPPUNER, R. (2004) Geographie und spatial turn, in: *Erdkunde* 58, p. 201–211.
LURY, C. (2004) *Brands: the logis of global economy*, Routledge: London.
LYNCH, M. (1995) Building a global infrastructure, in: *Studies in History and Philosophy of Science* 26, p. 167–172.
MA, K. W. (2002) Translocal spatiality, in: *International Journal of Cultural Studies* 5, p. 131–152.
MAGOLDA, P. (2000) Accessing, waiting, plunging in, wondering, and writing: Retrospective sense-making of fieldwork, in: *Field Methods* 12, p. 209–234.
MAHLER, S. & PESSAR, P. (2001) Gendered geographies of power: analysing gender across transnational spaces, in: *Identities* 7, p. 441–459.
MAINTZ, J. (2008) Synthesizing the face-to-face experience: E-learning practices and the constitution of place online, in: *Social Geography* 3, p. 1–10.
MALINOWSKI, B. (1922) *Argonauts of the western pacific*, London: Routledge.
MALKKI, L. (1992) National geographic: The rooting of peoples and the territorialization of national identity among scholars and refugees, in: *Cultural Anthropology* 7, p. 24–44.
MANDAVILLE, P. G. (2001) *Transnational Muslim politics: Reimagining the Umma*, Routledge: London.
MANSVELT, J. (2008) Geographies of consumption: citizenship, space and practice, in: *Progress in Human Geography* 32, p. 105–117.
— (2009) Geographies of consumption: the unmanageable consumer, in: *Progress in Human Geography* 32, p. 264–274.
— (2010) Geographies of consumption: engaging with absent presences, in: *Progress in Human Geography* 34, p. 224–233.
MARCUS, G. (1986) Contemporary problems of ethnography in the modern world system, in: Clifford, J. & Marcus, G. (eds.) *Writing culture*, University of California Press: Berkeley, p. 165–193.
— (1989) Imagining the whole: ethnography's contemporary efforts to situate itself, in: *Critical Anthropology* 9, p. 7–30.
— (1995) Ethnography in/of the world system: the emergence of multi-sited ethnography, in: *Annual Review of Anthropology* 24, p. 95–117.
— (2009) Beyond Malinowski and after 'writing culture': On the future of cultural anthropology and the predicament of ethnography in: *The Australian Journal of Anthropology* 13, p. 191–199.
MARCUS, G. & OKELY, J. (2007) How short can fieldwork be?, in: *Social Anthropology/Anthropologie Sociale* 15, 3, p. 353–367.
MARSTON, A., JONES III, J. P. & WOODWARD, K. (2005) Human geography without scale, in: *Transactions of the Institute of British Geographers* 30, p. 416–432.
MARTIN, R. & SUNLEY, P. (2001) Rethinking the "economic" in economic geography: Broadening our vision or losing our focus?, in: *Antipode* 33, p. 148–161.
MASSEY, D. (1991) A global sense of place, in: *Marxism Today* p. 24–29.
— (1994a) *Space, place, and gender*, University of Minnesota Press: Minneapolis.

—— (1994b) A global sense of place, in: Massey, D. (ed.) *Space, place and gender*, Polity Press: Cambridge, p. 146–156.
—— (2004a) The political challenge of relational space: Introduction to the Vega Symposium, in: *Geografiska Annaler B* 86, p. 3–4.
—— (2004b) Geographies of responsibility, in: *Geografiska Annaler B* 86, p. 5–18.
MATEOS, P. & FISHER, P. (2007) Spatiotemporal accuracy in mobile phone location: assessing the new cellular geography, in: Drummond, J., Billen, R., Forrest, D. and E. Joao (eds.) *Dynamic & Mobile GIS: Investigating Change in Space and Time*, Taylor & Francis: London, p. 189–212.
MAUSS, M. (1923/1924]) Essai sur le don. Forme et raison de l'échange dans les sociétés archaïque, in: *L'Année Sociologique* 1, p. 30–186.
MAY, J. (1996a) A little taste of something more exotic: the imaginative geographies of everyday life, in: *Geography Compass* 81, p. 57–64.
—— (1996b) Globalisation and the politics of place: place and identity in an inner London neighbourhood, in: *Transactions of the Institute of British Geographers* 21, p. 194–215.
MAZRUI, A. & SHARIIF, I. N. (1994) *The Swahili: idiom and identity of an African people*, Africa World Press: Trenton NY.
MCHUGH, K. (2000) Inside, outside, upside down, backward, forward, round and round: a case for ethnographic studies in migration, in: *Progress in Human Geography* 24, p. 71–89.
MCKAY, D. (2006) Translocal circulation: Place and subjectivity in an extended Filipino community, in: *The Asia Pacific Journal of Anthropology* 7, p. 265–278.
MEGORAN, N. (2006) For ethnography in political geography: experiencing and re-imagining Ferghana Valley boundary closures, in: *Political Geography* 25, p. 622–640.
MERRIMAN, P. (2007) *Driving Spaces: a cultural-historical geography of England's M1 motorway*, Blackwell: Oxford.
MIDDLETON, J. (1992) *The world of the Swahili. An African mercantile civilization*, Yale University Press: London.
—— (2003) Merchants: an essay in historical ethnography, in: *Journal of the Royal Anthropological Institute* 9, p. 509–526.
—— (2004) *African merchants of the Indian Ocean. Swahili of the East African coast*, Waveland Press: Long Grove.
MIDDLETON, J. (2011) Walking in the city: the geographies of everyday pedestrian practices, in: *Geography Compass* 5, p. 90–105.
MILLER, D. (1998) *A theory of shopping*, Polity Press: Cambridge.
MILLER, D. & SLATER, D. (2000) *The internet: An ethnographic approach*, Berg: Oxford.
MILLER, P. (1998) The multiplying machine, in: *Accounting, Organisation and Society* 22, p. 355–364.
MINTZ, S. (1986) *Sweetness and power: the place of sugar in modern history*, Penguin Books: New York.
MITCHELL, D. (1995) There is no such thing as culture: towards a reconceptualisation of the idea of culture in geography, in: *Transactions of the Institute of British Geographers* 20, p. 102–116.
—— (2000) *Cultural geography: A critical introduction*, Blackwell: Oxford.
MITCHELL, K. (1997) Transnational discourses: Bringing geography back in, in: *Antipode* 29, p. 101–114.
MOELLER, T. (1971) *Bergbau und regionale Entwicklung in Ostafrika*, Weltforum Verlag: Bonn.
Mol, A. & LAW, J. (1994) Regions, networks and fluids: Anaemia and social topology, in: *Social Studies of Science* 24, p. 641–671.
MORLEY, D. (2000) *Home territories: Media, mobility and identity*, Routledge: London.
MURDOCH, J. (1997) Towards a geography of heterogeneous associations, in: *Progress in Human Geography* 21, p. 321–337.
—— (2006) *Post-structuralist geography*, Sage: London.

MUTCH, A. (2002) Actors and networks or agents and structures: Towards a realist view of information systems, in: *Organization* 9, p. 477–496.
MYERS, G. (1993) Isle of cloves, sea of discourses: Writing about, in: *Cultural Geographies* 3, p. 408–426.
—— (1994) Making the socialist city of Zanzibar., in: *The Geographical Review* 84, p. 451–464.
—— (2000) Narrative representations of revolutionary Zanzibar, in: *Journal of Historical Geography* 26, p. 429–448.
NAGAR, R. (1997) Exploring methodological borderlands through oral narratives, in: Jones III., J. P., Nast, H. & Roberts, S. (eds.) *Thresholds in feminist geography*, Bowman & Littlefield Publishers: Plymouth, p. 203–224.
NAGAR, R. & ALI, F. (2003) Collaboration across borders: Moving beyond positionality, in: *Singapore Journal of Tropical Geography* 24, p. 356–372.
NAIPAUL, V. S. (1979) *A bend in the river*, Pan Macmillan: London.
NEWTON, I. (1687) *Philosophiae naturalis principia mathematica*.
NICHOLLS, C. S. (1971) *The Swahili coast – Politics, diplomacy and trade on the East African littoral 1798–1856*, Africana Publishing Corperation: New York.
NICHOLLS, W. (2009) Place, networks, space: Theorising the geographies of social movements, in: *Transactions of the Institute of British Geographers* 34, p. 78–93.
NICOLINI, B. (2004) *Makran, Oman, and Zanzibar: Three-terminal cultural corridor in the western Indian Ocean*, Brill: Leiden.
—— (2006) The Makran-Baluch-African network in Zanzibar and East Africa during the XIXth Century, in: *Journal of African and Asian Studies* 5, p. 347–370.
NOGUCHI, Y. (2004) Trying to stay in touch? Telecom companies are eager to help Hispanic families do just that, in: *Washington Post* 6. June, p. F08.
NORA, P. (1989) Between memory and history: Les lieux de mémoire, in: *Representations* 26, p. 7–25.
NURSE, D. & SPEAR, T. (1985) *The Swahili. Reconstructing the history and language of an African society, 800–15000*, University of Pennsylvania Press: Philadelphia.
NYAMJOH, F. (2008) Globalization and the Cultural Economy: Africa, in: Anheier, H. & Isar, Y. (eds.) *Cultures and Globalization. The Cultural Economy*. Sage: London, p. 123–134.
NYMAN, J. (2009) *Home, identity, and mobility in contemporary diasporic fiction*, Rodopi: Amsterdam.
OAKES, T. & SCHEIN, L. (2006) Translocal China. An Introduction, in: *Translocal China. Linkages, identities and the reimagining of space*, Routledge: London, p. 1–35.
OGBORN, M. (2002) Writing travels: Power, knowledge and ritual on the English East India Company´s early voyages, in: *Transactions of the Institute of British Geographers* 27, p. 155–171.
OKELLO, J. (1971) *The Zanzibar revolution*, East African Literature Bureau: Nairobi.
ONG, A. & NONINI, D. (1997) *Ungrounded empires: The cultural politics of modern Chinese transnationalism*, Routledge: New York.
PALLASMAA, J. (2008) Existential homelessness – Placelessness and nostalgia, in: Bergmann, S. & Sager, T. (eds.) *The ethics of mobilities: rethinking place, exclusion, freedom and environment*, Ashgate: London, p. 143–156.
PANAGAKOS, A. N. & HORST, H. A. (2006) Return to Cyberia: Technology and the social worlds of transnational migrants, in: *Global Networks* 6, p. 109–124.
PARKIN, D. (1994) *Continuity and autonomy in Swahili communities*, Afro-Pub: Vienna.
PASCUAL-DE-SENS, A. (2004) Sense of place and migration histories – idiotopy and idiotope, in: *Area* 36, p. 348–357.
PEARCE, F. (1920) *Zanzibar: The island metropolis of Eastern Africa*, T. Fisher Unwin: London.
PEARSON, M. (2003) *The Indian Ocean*, Routledge: London.
PELS, D., HETHERINGTON, K. & Vandenberghe, F. (2002) The status of the object: Performances, mediations, and techniques, in: *Theory, Culture and Society* 19 (5–6), p. 1–21.

PELTONEN, T. (2007) In the middle of managers. Occupational communities, global ethnography and the multinationals, in: *Ethnography* 8, p. 346–360.
PETERSON, J. E. (2004) Oman's diverse society: Northern Oman, in: *Middle East Journal* 58, p. 32–51.
PETRIDOU, E. (2001) The taste of home, in: Miller, D. (ed.) *Home possessions. Material culture behind closed doors*, Berg: Oxford, p. 87–104
PHILO, C. (1991) *New words, new worlds: reconceptualising social and cultural geography*, Cambrian Printers: Aberystwyth.
PIETERSE, J. N. (2003) *Global Melange: Globalisation and culture*, Rowman and Littlefield: Lanham.
PILE, S. & THRIFT, N. (1995) *Mapping the subject: Geographies of cultural transformation*, Routledge: London.
PINK, S. (2007) *Doing visual ethnography: Images, media and representation in research*, Sage: London.
—— (2009) *Doing sensory ethnography*, Sage: London.
PLATTNER, S. (1989) Economic behaviour in markets, in: Plattner, S. (ed.) *Economic anthropology*, Stanford University Press: Paolo Alto, p. 209–222.
PODOLNY, J. M. & BARON, J. N. (1997) Relationships and resources: Social networks and mobility in the workplace, in: *American Sociological Review* 62, p. 673–693.
POLANYI, K. (1944) *The great transformation. The political and economic origins of our time*, Farrar & Rinehart: New York.
POPP, H. (1979) Zur Bedeutung des Koppelungsverhaltens bei Einkäufen in Verbrauchermärkten: aktionsräumliche Aspekte, in: *Geographische Zeitschrift* 67, p. 301–313.
—— (1993) Kulturgeographie ohne Kultur?, in: Hansen, K. (ed.) *Kulturbegriff und Methode: der stille Paradigmenwechsel in den Geisteswissenschaften*, Gunter Narr Verlag: Tübingen, p. 115–132.
PORTES, A. (2003) Conclusion: Theoretical convergencies and empirical evidence in the study of immigrant transnationalism, in: *International Migration Review* 37, p. 874–892.
PORTES, A. & DEWIND, J. (2007) *Rethinking migration: New theoretical and empirical perspectives*, Berghahn Books: New York.
PORTES, A., GUARNIZO, L. & LANDOLT, P. (1999) The study of transnationalism: Pitfalls and promise of an emergent research field, in: *Ethnic and Racial Studies* 22, p. 217–237.
POUWELS, R. (1991) Swahili networks, in: *Cahiers d'Études africaines* 123, p. 411–415.
POWELL, R. (2007) Geographies of science: Histories, localities, practices, futures, in: *Progress in Human Geography* 31, p. 309–329.
PRED, A. (1977) The choreography of existence: Some comments on Hägerstrand's time geography and its effectiveness, in: *Economic Geography* 53, p. 207–221.
PRESTHOLDT, J. (2009) Mirroring modernity: On consumerism in cosmopolitan Zanzibar, in: *Transforming Cultures* 4, p. 165–204.
PRIES, L. (1996) Transnationale soziale Räume. Theoretisch-empirische Skizze am Beispiel der Arbeitswanderungen Mexico-USA, in: *Zeitschrift für Soziologie* 25, p. 456–472.
—— (1999a) New migration in transnational spaces, in: Pries, L. (ed.) *Migration and transnational social spaces*, Ashgate: Aldershot, p. 1–35.
—— (1999b) Die Transnationalisierung der sozialen Welt und die deutsche Soziologie, in: *Soziale Welt* 50, p. 383–394.
—— (2001a) The disruption of social and geographic space, in: *International Sociology* 16, p. 51–70.
—— (2001b) The approach of transnational social spaces: Responding to new configuration of the social and spatial, in: Pries, L. (ed.) *New transnational social spaces. International migration and transnational companies*, London: Routledge, p. 3–33.
—— (2008) *Die Transnationalisierung der sozialen Welt*, Suhrkamp: Frankfurt a.M.

PRINS, A. H. J. (1961) *The Swahili-speaking peoples of Zanzibar and the East African coast*, International African Institute: London.
—— (1965) *Sailing from Lamu: A study of maritime culture in Islamic East-Africa*, Van Gorcum: Assen.
RAAB, K. (2006) *Rapping the nation*, LIT Verlag: Münster.
RABINOW, P. (1977) *Reflections on fieldwork in Marocco*, University of California Press: Berkeley.
RAJCHMAN, J. (2000) *The Deleuze connections*, MIT Press: Cambridge, MA.
RAY, L. & SAYER, A. (1999) *Culture and economy after the cultural turn*, Sage: London.
REDEPENNING, M. (2007) Eine Ästhetik der Unverbindlichkeit? Kultur als jüngere Selbstbeschreibung der Geographie, in: Berndt, C. & Pütz, R. (eds.) *Kulturelle Geographien: zur Beschäftigung mit Raum und Ort nach dem Cultural Turn*, Transcript: Bielefeld, p. 349–378.
—— (2008) Eine selbst erzeugte Überraschung: Zur Renaissance von Raum als Selbstbeschreibungsformel der Gesellschaft, in: Döring, J. & Thielmann., T. (eds.) *Spatial turn. Das Raumparadigma in den Kultur- und Sozialwissenschaften*, Transcript: Bielefeld, p. 317–340.
REICHARD, P. (1892) Deutsch-Ostafrika: das Land und seine Bewohner, seine politische und wirtschaftliche Entwicklung, Spamer: Leipzig.
REICHENBACH, A. (2005) 160 Zeichen Liebe und Subversion, in: Geisenhainer, K. & Lange, K. (eds.) *Bewegliche Horizonte*, Leipziger Universitätsverlag: Leipzig, p. 351–376.
RELPH, E. (1976) *Place and Placelessness*, Pion: London.
REMES, P. (1998) *Karibu ghetto langu - Welcome in my ghetto: Urban youth, language and culture in 1990s Tanzania*, PhD-thesis, Northwestern University: Chicago.
REMUS, G. A., GERHARZ, E. & RESCHER, G. (2008) *The making of world society: Perspectives from transnational research*, Transcript: Bielefeld.
RENAULT, F. (1987) *Tippo Tip - un potentat arabe en Afrique centrale au XIXe siècle*, L´Harmattan: Paris.
RETTIE, R. (2008) Mobile phones as network capital: Facilitating connections, in: *Mobilities* 3, p. 291–311.
REUBER, P. & PFAFFENBACH, C. (2005) *Methoden der empirischen Humangeographie*, Westermann: Braunschweig.
REUSTER-JAHN, U. & KIESSLING, R. (2006) *Lugha Ya Mitaani* in Tanzania. The poetics and sociology of a young urban style of speaking, in: *Swahili Forum* 13, p. 1–200.
RICKETTS HEIN, J., EVANS, J. & JONES, P. (2008) Mobile methodologies: Theory, technology and practice, in: *Geography Compass* 2, p. 1266–1285.
ROBERTSON, R. (1995) Glocalisation: Time-space and heterogeneity-homogeneity, in: Featherstone, M., Lash, S. & Roberston, R. (eds.) *Global modernities*, Sage: London, p. 25–44.
ROLDAN, A. (1995) Malinowski and the origins of the ethnographic method, in: Roldan, A. & Vermeulen, H.(eds.) *Fieldwork and footnotes. Studies in the history of European anthropology*, Routledge: London, p. 143–158.
ROSE, G. (1997) Situating knowledges: Positionality, reflexivities and other tactics, in: *Progress in Human Geography* 21, p. 305–320.
—— (2007) *Visual methodologies: An introduction to the interpretation of visual materials*, Sage: London.
ROTTENBURG, R., KALTHOFF, H. & WAGENER, H.-J. (2000) Introduction: In search of a new bed: Economic representations and practices, in: Kalthoff, H., Rottenburg, R. & Wagener, H.-J. (eds.) *Facts and figures – Economic representations and practices*, Metropolis Verlag: Marburg, p. 9–34.
SAID, E. (1993) *Culture and imperialism*, New York.
—— (2003 [1978]) *Orientalism*, Penguin Books: London.
SALEH, M. A. (2002) Tolerance. The principle foundation of the cosmopolitan society of Zanzibar in: *Journal of Cultures of the World*, UNESCO Centre of Catalunya.

—— (2004) Going with the times: Swahili norms and values today, in: Caplan, P. & Topan, F. (eds.) *Swahili modernities*, Africa World Press: Trenton, NJ, p. 145–155.

—— (2006) Les investissements des Zanzibari à Kariakoo, in: Calas, B. (ed.) *De Dar es Salaam a Bongoland. Mutations urbaines en Tanzanie*, Karthala: Paris, p. 353–368.

SALIM, A. I. (1973) *The Swahili-speaking peoples of Kenya's coast, 1895–1945*, East African Publishing House: Nairobi.

SANDERS, R. (2008) The triumph of geography, in: *Progress in Human Geography* 32, p. 179–182.

SASSEN, S. (1991) *The global city*, Princeton: New York.

SAUER, C. (1925) The morphology of landscape, in: *University of California Publications in Geography* 2, p. 19–54.

SAYER, A. (1991) Behind the locality debate: Deconstructing geography's dualisms, in: *Environment and Planning A* 23, p. 283–308.

—— (1997) The dialectic of culture and economy, in: Lee, R. & Willis, J. (eds.) *Geographies of economies*, Arnold: London, p. 16–26.

SCHLOTTMANN, A. (2008) Closed spaces: Can't live with them, can't live without them, in: *Environment and Planning D: Society and Space* 26, p. 823–841.

SCHROER, M. (2006) *Räume, Orte, Grenzen. Auf dem Weg zu einer Soziologie des Raums*, Suhrkamp: Frankfurt a.M.

SCHWEIZER, T. (1993) Perspektiven der analytischen Ethnologie, in: Schweizer, T., Schweizer, M. & Kokot, W. (eds.) *Handbuch der Ethnologie*, Dietrich Reimer Verlag: Berlin, p. 79–113.

SCOTT, J. (2000) *Social network analysis: a handbook*, Sage: London.

SEAMON, D. (1979) *Geography of the lifeworld: Movement, rest, and the encounter*, St. Martin's Press: New York.

SERRES, M. (1991) *Hermes 1. Kommunikation*, Merve Verlag: Berlin.

SERRES, M. & LATOUR, B. (1995) *Conversations on science, culture and time*, University of Michigan Press:

SHAW, C. & CHASE, M. (1989) *The imagined past: History and nostalgia*, Manchester University Press: Manchester.

SHELLER, M. (2004) Mobile publics: Beyond the network perspective, in: *Environment and Planning D: Society and Space* 22, p. 39–52.

—— (2008) Gendered mobilities: Epilogue, in: Uteng, T. P. & Cresswell, T. (eds.) *Gendered Mobilities*, Ashgate: London, p. 257–266.

SHELLER, M. & URRY, J. (2006) The new mobilities paradigm, in: *Environment and Planning A* 38, p. 207–226.

SHERIFF, A. (1971) The rise of a commercial empire: an aspect of the economic history of Zanzibar, unpublished PhD-thesis, SOAS, University of London.

—— (1995) *The history and conservation of Zanzibar Stone Town*, James Currey: Oxford.

—— (2001) Capitalist Hegemony Reconsidered, in: Marfaing, L. & Reinwald, B. (eds.) *African networks, exchange and spatial dynamics*, Lit: Hamburg, p. 147–154.

—— (2006) Between the worlds: The littoral peoples of the Indian Ocean, in: Loimeier, R. & Seesemann, R. (eds.) *The global worlds of the Swahili*, LIT: Münster, p. 15–30.

—— (2008) The dhow culture of the Western Indian Ocean, in: Basu, H. (ed.) *Journeys and dwellings*, Orient Longman: London and Hyderabad, p. 61–89.

—— (2009) The Persian Gulf and the Swahili coast: A History of acculturation over the longue durée, in: Potter, L. (ed.) *The Persian Gulf in history*, Palgrave Macmillan: Basingstoke, p. 173–188.

SHIELDS, R. (1992) *Lifestyle shopping: The subject of consumption*, Routledge: London.

SIMMEL, G. (1890) *Über sociale Differenzierung*, Duncker & Humblot: Leipzig.

—— (1908) *Soziologie. Untersuchungen über die Formen der Vergesellschaftung*, Duncker & Humblot: Leipzig.

SIMPSON, E. & KRESSE, K. (2008) Introduction. Cosmopolitanism contested: Anthropology and history in the Western Indian Ocean, in: Simpson, E. & Kresse, K. (eds.) *Struggling with history. Islam and cosmopolitanism in the Western Indian Ocean*, Columbia University Press: New York, p. 1–41.

SLATER, D. (1997) *Consumer culture and modernity*, Polity Press: Cambridge.

—— (2002) Capturing markets from the economists, in: DuGay, P. & Pryke, M. (eds.) *Cultural economy – cultural analysis and commercial life*, Sage: London, p. 59–77.

SMITH, M. P. (2001) *Transnational urbanism*, Blackwell: Oxford.

—— (2005) Transnational urbanism revisited, in: *Journal of Ethnic and Migration Studies* 31, p. 235–244.

SMITH, M. P. & EADE, J. (2008) *Transnational ties: Cities, migrations, and identities*, Transaction Publishers: New Brunswick.

SMITH, M. P. & GUARNIZO, L. (1998) *Transnationalism from below*, Transaction Publishers: New Brunswick.

SMITH, N. & KATZ, C. (1993) Grounding metaphor: towards a spatialized politics, in: Keith, M. & Pile, S. (eds.) *Place and the Politics of Identity*, Routledge: London, p. 67–83.

SMITH, R. (2003a) World city actor-networks, in: *Progress in Human Geography* 27, p. 25–44.

—— (2003b) World city topologies, in: *Progress in Human Geography* 27, p. 561–582.

SMOREDA, Z. & THOMAS, F. (2001) Use of SMS in Europe, in: *Usages* 1.

SOJA, E. (1989) *Postmodern geographies. The reassertion of space in critical social theory*, Verso: New York.

—— (1996) *Thirdspace: journeys to Los Angeles and other real-and-imagined places*, Blackwell: Oxford.

—— (2008) Vom 'Zeitgeist' zum 'Raumgeist'. New twists on the spatial turn, in: Döring, J. &. Thielmann, T. (eds.) *Spatial turn. Das Raumparadigma in den Kultur- und Sozialwissenschaften*, Transcript: Bielefeld, p. 241–262.

—— (2009) Taking space personally, in: Warf, B. & Arias, S. (eds.) *The spatial turn. Interdisciplinary perspectives*, Routledge: London, p. 11–35.

SPEKE, J. H. (1864) *Journal of the discovery of the source of the Nile*, Harper: New York.

SPITTLER, G. (1998) *Hirtenarbeit. Die Welt der Kamelhirten und Ziegenhirtinnen von Timia*, Rüdiger Köppe Verlag: Köln.

STÄHELI, U. (2000) *Poststrukturalistische Soziologien*, Transcript: Bielefeld.

STANLEY, H. (2005 [1909]) *The autobiography of Sir Henry Morton Stanley*, Kessinger Publications: Whitefish.

STEGBAUER, C. & HÄUSSLING, R. (2010) *Handbuch der Netzwerkforschung*, VS Verlag: Wiesbaden.

STELLRECHT, I. (1993) Interpretative Ethnologie, in: Schweizer, T., Schweizer, M. & Kokot, W. (eds.) *Handbuch der Ethnologie*, Dietrich Reimer Verlag: Berlin, p. 29–78.

STOCKING, G. (1982) *Race, culture, and evolution: essays in the history of anthropology*, University of Chicago Press: Chicago.

—— (2001) The shaping of national anthropologies. A view from the center, in: Stocking, G. (ed.) *Delimiting anthropology: occasional essays and reflections*, University of Wisconsin Press: Madison, p. 281–303.

STOLLER, P. (2002) *Money has no smell*, University of Chicago Press: Chicago.

—— (2003) African/Asian/Uptown/Downtown, in: Amin, A. & Thrift, N. (eds.) *The Blackwell cultural economy reader*, Wiley-Blackwell: Oxford, p. 193–209.

STROBEL, M. (1975) Women's wedding celebrations in Mombasa, Kenya, in: *African Studies Review* 18, p. 35–45.

SUNLEY, P. (2008) Relational economic geography: A partial understanding or a new paradigm, in: *Economic Geography* 84, p. 1–26.

SUTTON, J. E. G. (1970) *Dar es Salaam. City, port and region*, Tanzanian Society. Dar es Salaam.

SWEDBERG, R. & GRANOVETTER, M. (2001) *The sociology of economic life*, Westview Press: Boulder, Colorado.

SWYNGEDOUW, E. (1997) Neither global nor local: 'Glocalisation' and the politics of scale, in: Cox, K. (ed.) *Spaces of globalisation: reasserting the power of the local*, Guilford: New York, p. 137–166.

TAYLOR, P. (2004) *World city network: A global urban analysis*, Routledge: London.

TAYLOR, P. et al. (2001) A new mapping of the world for the new millennium, in: *The Geographical Journal* 167, p. 213–222.

THANI, A. (2003) Amani Thani Fairoz, in: Barwani, S. et al. (eds.) *Unser Leben vor der Revolution und danach – maisha yetu kabla ya mapinduzi na baadaye*, Rüdiger Köppe Verlag: Köln, p. 135–267.

THORNTON, P., WILLIAMS, A. & SHAW, G. (1997) Revisiting time-space diaries: An exploratory case study of tourist behaviour in Cornwall, England, in: *Environment and Planning A* 29, p. 1847–1867.

THRIFT, N. (1995) Inhuman geographies: Landscapes of speed, light, and power, in: Cloke, P. et al. (eds.) *Writing the Rural: Five Cultural Geographies*, SAGE: London, p. 191–248.

—— (2000a) Actor-Network-Theory, in: Gregory, D. et al. (eds.) *Dictionary of Human Geography*, Blackwell: Oxford, p. 716–717.

—— (2000b) Pandora's box? Cultural geographies of economies, in: Clark, G., Feldmann, M. & Gertler, M. (eds.) *The Oxford handbook of economic geography*, Oxford University Press: Oxford, p. 689–702.

—— (2000c) Dead or alive?, in: Cook, I. et al. (eds.) *Cultural turns /Geographical turns*, Prentice Hall: Harlow, p. 1–6.

THRIFT, N. & PRED, A. (1981) Time geography: A new beginning, in: *Progress in Human Geography* 5, p. 277–286.

TOLIA-KELLY, D. (2004) Locating processes of identification: Studying the precipitates of re-memory through artefacts in British Asian home, in: *Transactions of the Institute of British Geographers* 29, p. 314–329.

TOPAN, F. (1998) Langue et culture swahili à Zanzibar, in: Le Cour Grandmaison, C. & Crozon, A. (eds.) *Zanzibar aujourd'hui*, Karthala: Paris, p. 246–257.

—— (2000) Being a Muslim in East Africa: A Swahili perspective, in: Salter, T. & King, K. (eds.) *Africa, Islam and development*, Centre of African Studies: Edinburgh, p. 283–298.

—— (2006) From coastal to global: The erosion of the Swahili 'Paradox', in: Loimeier, R. & Seesemann, R. (eds.) *The global worlds of the Swahili*, LIT: Münster, p. 55–66.

TOWNSEND, J. (1977) *Oman: The making of a modern state*, Helm: London.

TSING, A. L. (2005) *Friction: an ethnography of global connection*, Princeton University Press: Princeton.

TUAN, Y. F. (1974) *Topophilia: A study of environmental perception, attitudes and values*, Prentice Hall: Eaglewood Cliffs.

—— (1978) Sign and metaphor, in: *Annals of the Association of American Geographers* 68, p.363–372.

—— (1991) Language and the making of place: A narrative-descriptive approach, in: *Annals of the Association of American Geographers* 81, p. 684–696.

—— (2001a) Life as a fieldtrip, in: *Geographical Review* 91, p. 41–45.

—— (2001b [1977]) *Space and Place: The perspective of experience*, University of Minnesota Press: Minneapolis.

TUFTE, T. (2002) Ethnic minority Danes between diaspora and locality – Social users of mobile phone and internet, in: Stald, G. & Tufte, T. (eds.) *Global encounters: Media and cultural transformation*, University of Luton Press: Luton, p. 235–262.

URRY, J. (2007) *Mobilities*, Polity Press: Cambridge.

VALCKE, S. (1999) Entrepreneurs: Business à Zanzibar, in: Baroin, C. & Constantin, F. (eds.) *La Tanzanie contemporaine*, Karthala: Paris, p. 335–347.

VALENTINE, G. (2001) *Social Geographies – Space and society*, Person Educational Limited: Essex.
VALERI, M. (2007) Nation-building and communities in Oman since 1970: the Swahili-speaking Omani in search for identity, in: *African Affairs* 106, p. 497–506.
VASSANJI, M. G. (1991) *Uhuru Street*, Heinemann International: London.
VERNE, J. & DOEVENSPECK, M. (2012) Bitte da bleiben! Sedentarismus als Konstante in der Migrationsforschung, in: Steinbrink, M. & Geiger, M. (eds.) *Migration und Entwicklung aus geographischer Perspektive*, IMIS, Osnabrück.
VERNE, J. & MÜLLER-MAHN, D. (2012) 'We are part of Zanzibar' – Translocal practices and imaginative geographies in contemporary Oman-Zanzibar relations, in: Wippel, S. (ed.) *Regionalising Oman*, Springer Science: Dordrecht, Netherlands.
VERTOVEC, S. (1999) Conceiving and researching transnationalism, in: *Ethnic and Racial Studies* 22, p. 447–462.
—— (2009) *Transnationalism*, Routledge: London.
VIERKE, C. (2011) *On the Poetics of the Utendi. A Critical Edition of the Nineteenth-Century Swahili Poem 'Utendi wa Haudaji' together with a Stylistic Analysis*, LIT: Münster,
VILLIERS, I. A. (1948) Some aspects of the Arab dhow trade, in: *Middle East Journal* 2, p. 400–404.
—— (1952) *Monsoon seas*, McGraw-Hill Book Company: New York.
VOIGT-GRAF, C. (2004) Towards a geography of transnational spaces: Indian transnational communities in Australia, in: *Global Networks* 4, p. 25–49.
WALKER, I. (2008) Hadramis, Shimalis and Muwalladin: Negotiating cosmopolitan identities between the Swahili coast and southern Yemen, in: *Journal of Eastern African Studies* 2, p. 44–59.
—— (2011) Hybridity, belonging, and mobilities: The intercontinental peripatetics of a transnational community, in: *Population, Space and Place* 17, p. 167–179.
WALSH, K. (2006) British expatriate belongings: Mobile homes and transnational homing, in: *Home cultures* 3, p. 123–144.
—— (2009) Geographies of the heart in transnational spaces: Love and the intimate lives of British migrants in Dubai, in: *Mobilities* 4, p. 427–445.
WARF, B. & ARIAS, S. (2009) *The spatial turn. Interdisciplinary perspectives*. Routledge: London.
WARNER WOOD, W. (2001) Rapport is overrated: Southwestern ethnic art dealers and ethnographers in the 'field', in: *Qualitative Inquiry* 7, p. 484–503.
WASSERMANN, S. & FAUST, K. (1994) *Social network analysis*, Cambridge University Press: Cambridge.
WATERS, J. (2010) Becoming a father, missing a wife: Chinese transnational families and the male experience of lone parenting in Canada, in: *Population, Space and Place* 16, p. 63–74.
WATTS, L. & URRY, J. (2008) Moving methods, travelling times, in: *Environment and Planning D: Society and Space* 26, p. 860–874.
WEIGEL, S. (2002) Zum 'topographical turn'. Kartographie, Topographie und Raumkonzepte in den Kulturwissenschaften, in: *KulturPoetik* 2, p. 151–165.
WERLEN, B. (1995) *Sozialgeographie alltäglicher Regionalisierungen, Bd.1: Zur Ontologie von Gesellschaft und Raum*, Franz Steiner Verlag: Stuttgart.
—— (1997) *Sozialgeographie alltäglicher Regionalisierungen, Bd. 2: Globalisierung, Region, Regionalisierung*, Franz Steiner Verlag: Stuttgart.
—— (2008) Körper, Raum und mediale Repräsentationen, in: Döring, J. & Thielmann, T. (eds.) *Spatial turn. Das Raumparadigma in den Kultur- und Sozialwissenschaften*, Transcript: Bielefeld, p. 365–392.
WHATMORE, S. (1999) Hybrid geographies: Rethinking the `human' in human geography, in: Massey, D., Allen, J., Sarre, P. (eds.) *Human geography today*, Polity Press: Cambridge, p. 24–39.

—— (2003) Generating materials, in: Pryke, M., Rose, G. & Whatmore, S. (eds.) *Using social theory: Thinking through research*, SAGE: London, p. 89–104.
WILES, J. (2008) Sense of home in a transnational social space: New Zealanders in London, in: *Global Networks* 8, p. 116–137.
WILLIAMS, R. (1976) *Keywords: A vocabulary of culture and society*, Fontana: London.
WILLIS, J. (1993) *Mombasa, the Swahili, and the Making of the Mijikenda*, Hurst: London.
WILLIS, K., YEOH, B. & ABDUL KHADER FAKHRI, S. M. (2004) Introduction. Transnationalism as a challenge to the nation, in: Yeoh, B. & Willis, K. (eds.) *State/Nation/Transnation*, Routledge: London, p. 1–15.
WILLIS, R. G. (1981) *A state in the making: Myth, history and social transformation in precolonial Ufipa*, Indiana University Press: Bloomington.
WIMMER, A. & GLICK SCHILLER, N. (2002) Methodological nationalism and beyond: Nation-state building, migration and the social sciences, in: *Global Networks* 2, p. 301–334.
WIRTH, E. (1979) *Theoretische Geographie. Grundzüge einer theoretischen Kulturgeographie*, B.G. Teubner: Stuttgart.
WOGAN, P. (2004) Deep hanging out: Reflections on fieldwork and multi-sited Andean ethnography, in: *Identities* 11, p. 129–139.
WULFF, H. (2002) Yo-yo fieldwork: Mobility and time in a multilocal study of dance in Ireland, in: *Anthropological Journal on European Cultures* 11, p. 117–136.
WYLIE, J. (2005) A single day's walking: narrating self and landscape on the South West coast, in: *Transactions of the Institute of British Geographers* 30, p. 234–247.
YEUNG, H. W.-C. (1994) Critical reviews of geographical perspectives on business organisations and the organisation of production: Towards a network approach, in: *Progress in Human Geography* 18, p. 460–490.
—— (1998) *Transnational corporations and business networks: Hong Kong firms in the ASEAN region*, Routledge: London.
—— (2005) Rethinking relational economic geography, in: *Transactions of the Institute of British Geographers* 30, p. 37–51.
ZELINSKY, W. (1973) *The cultural geography of the United States*, Prentice-Hall: Eaglewood Cliffs.
ZERUBAVEL, E. (1991) *The fine line: Making distinctions in everyday life*, Free Press: Detroit.
ZHOU, M. (2008) Revisiting ethnic entrepreneurship. Convergencies, controversies and conceptual advancements, in: Portes, A. & DeWind, J. (eds.) *Rethinking migration: New theoretical and empirical perspectives*, Berghahn Books: New York, p. 219–254.
ZHOU, Y. & TSENG, Y. F. (2001) Regrounding the 'ungrounded empires': Localization as the geographical catalyst for transnationalism, in: *Global Networks* 1, p. 131–154.